CROP PROTECTION MONOGRAPHS

Editor-in-Chief
J. Kranz, Gießen, Fed. Rep. of Germany

Editorial Board
K. H. Büchel, Leverkusen, Fed. Rep. of Germany
R. E. Frisbie, College Station, TX, USA
J. Palti, Tel Aviv, Israel
R. L. Zimdahl, Fort Collins, CO, USA

CROP PROTECTION MONOGRAPHS

Already published volumes

J. Palti, R. Ausher (Eds.)
Advisory Work in Crop Pest and Disease Management

H. Waibel
The Economics of Integrated Pest Control in Irrigated Rice

D. P. Singh
Breeding for Resistance to Diseases and Insect Pests

Further volumes in preparation

I. Wahl and G. Fischbeck
Role of Wild Relatives in Control of Crop Diseases,
Development of Their Epidemics, and Evolution of Pathogens

Dhan Pal Singh

Breeding for Resistance to Diseases and Insect Pests

With 19 Figures and 28 Tables

Springer-Verlag
Berlin Heidelberg New York
London Paris Tokyo

Dr. DHAN PAL SINGH
Department of Plant Breeding
Govind Ballabh Pant
University of Agriculture
and Technology
Pantnagar 263145
India

ISBN-13: 978-3-642-71514-3   e-ISBN-13: 978-3-642-71512-9
DOI: 10.1007/978-3-642-71512-9

To my father
Sri Praveen Singh
and to my mother
Smt. Dropadi Devi

Preface

The object of this book is to provide insight into the principles of disease and insect-pest resistance and to elaborate the resistance breeding practices with specific examples from as many different crops and parasites as possible. It is assumed that the readers are already in possession of some knowledge of plant pathogens and insect pests and their genetics from standard courses and text books. The book can be used for teaching an advanced course on the subject, such as in university lectures to graduate students. In addition, it should be useful as a reference book to plant pathologists, entomologists and plant breeders engaged in developing varieties resistant to harmful parasites.

I wish to express my sincere thanks to Dr. B.D. Singh, Banaras Hindu University, Varanasi, India, Dr. D. Sharma and Dr. S. Dwivedi, ICRISAT, Hyderabad, India; Dr. I.S. Singh and Dr. A.K. Bhattacharya, G.B. Pant University of Agriculture and Technology, Pantnagar, India, who made comments on some sections of the book.

Thanks are also due to Dr. D.N. Chaudhary, Dr. R.P.S. Verma and Mr. K.R. Reddy, who have given valuable help in one way or another in the publication of this book. I express my sincere thanks to Professor J.S. Nanda, Ex-Professor Plant Breeding in G.B.Pant University of Agriculture and Technology, Pantnagar, India for inspiring me to write this book. However, responsibility for errors and misinterpretations is entirely mine.

I would be remiss if I did not also acknowledge my debt to my wife Asha, and to my children Shailendra, Kanchan and Arti for their encouragement and patience during the preparation of this work.

The readers are requested to comment and give suggestions for further improvement in the book.

Pantnagar, Summer 1986   D.P. Singh

Contents

1 The Value of Disease and Insect Pest Resistance

A plant is healthy or normal when it carries out its physiological functions to the best of its genetic potential. Whenever plants are disturbed by pathogens or insect pests, one or more of the physiological functions is interfered with beyond a certain deviation from the normal.

Nearly all diseases/pests reduce biomass and hence yield. This may happen in either of the four ways: (1) by killing of plants, leaving a gap in the stand beyond the capacity of neighbours to compensate (e.g. vascular wilts, various soil borne fungi, some boring insects); (2) by general stunting caused by metabolic distruption, nutrient drain or root damage (e.g. many viruses, aphids, eelworms); (3) by killing branches (e.g. many boring insects, some fungal diebacks); and (4) by destruction of leaf tissues (e.g. many rusts, mildews, blights, leaf spots and some burning insects). Other effects of disease/insect attacks are more general. These include: damage to crop product initiated in the field, but often more apparent after harvest (e.g. cereal smuts, various rots and borers of fruits and tubers); and effects on quality (e.g. insect or fungal blemishes of fruits and tubers).

Some of the major famines or food losses of crop plants associated with pest and disease epidemics in the past are:

- the Irish famine of the 1840's, due to potato leaf blight epidemic;
- the wheatless days of 1917 in the USA, due to stem rust epidemics;
- the Bengal famine in India of 1943 associated with the *(Drechslera oryzae)* brown spot disease of rice;
- the devastation of all *Victoria*-derived oats in the mid 1940's in the USA due to a fungus causing *Victoria* blight disease;
- the rapid shift from brown planthopper biotype 1 to biotype 2 during 1974 to 1976, when large areas in the Philippines and Indonesia were planted to a few semi-dwarf varieties of rice;
- the downy mildew epidemic in pearl millet caused by *Sclerospora graminicola* in India in 1973; and
- the severe out-break of bacterial blight (*Xanthomonas campestris* pv. *oryzae*) in Punjab state of India occurred in 1980 (A.P.K. Reddy 1980). The widely grown varieties IR8, Jaya and PR 106 were affected and several thousand hectares of blighted fields had to be ploughed by the farmers.

The optimum conditions for a disease to occur and develop are a combination of three factors (Fig. 1.1): susceptible host, infective pathogen and favourable environ-

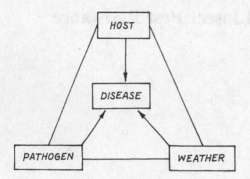

mental conditions. A change in any of the factors causes a corresponding change in the expression of disease.

Natural populations and wild plants rarely show epidemics, due to genetical heterogeneity and natural biological control (e.g. hyperparasitism). Hosts and parasites live together in a complex equilibrium in which neither dominates. In traditional agriculture, disease and insect epidemics were rare. However, modern agricultural technology has introduced important changes: (1) it has narrowed down the genetic base of cultivars, which alters the dynamic imbalance between host and parasites, which in turn results in epidemics; and (2) it has generated more or less continuously distributed populations and has changed the whole ecosystem, creating habitats profoundly altered for both host and parasite. This does not mean that all parasites become epidemic; they obviously do not and, indeed, the majority remain what they were before, unimportant. However, some of the minor parasites may increase due to favourable environment, host susceptibility (due to more virulent race or biotype) and continuity in large populations. For instance *Drechslera (maydis)*, leaf blight in maize in the USA and hoppers in rice, in India and the Philippines became major pests due to change in one or more of the above factors. The first dwarf variety of rice, Taichung Native-1, was introduced in 1964–65 for cultivation in India. It had resistance to blast *(Pyricularia oryzae)* but was susceptible to all other major pests and diseases. Meanwhile, bacterial leaf blight, *Xanthomonas campestris* pv. *oryzae*, which was localized, developed into an epidemic in 1963 on a local variety BR34 in Bihar. The cultivation of Taichung Native 1 was extended to many parts of India because of its high yields, in spite of its susceptibility to the bacterial blight in the country. This resulted in several epidemics (Safeeulla 1977). This shows that a disease, which was unknown or confined to small areas, reached epidemic proportions within a short period of 2 years (D.N. Srivastava 1972).

The prevention of epidemics and ultimately the reduction of losses in yield has been of great concern. The insect pests and some of the pathogens may be controlled by the use of chemicals; but the chemicals can create hazards to human health and can produce undesirable side effects on non-target insects, animals and plants. Therefore, development of resistant varieties is the most effective method of controlling many diseases and insect pests of crops such as cereal grains, which are grown on extensive acreages with a large number of plants per hectare.

Theophrastus, in the third century B.C., noted that cultivated varieties differed in their ability to avoid diseases. Andrew Knight (1799) and Darwin (1868) and several others in the middle of nineteenth century observed that crop varieties differed in disease resistance. Biffen (1905) showed that resistance to yellow rust in wheat was genetic and governed by a recessive gene. Stakman (1968) pointed out that wheat rusts, head blights, many leaf diseases, root rots and virus diseases cannot be controlled effectively and economically on a large scale without the help of resistant varieties, even though cultural practices may help in some areas. Similarly, resistant varieties are needed to control some of the most destructive disease/insect pests of rice, maize and sorghum etc.

Luginbill (1969) documented the utility of resistant varieties and reported that millions of dollars can be saved by growers. In the case of lettuce root aphid, the expression of resistance is unusually extreme and it endows the Wellesbourne cultivars with a freedom from root aphid attack superior to that obtained at extra cost and effort by the application of diazinon to the susceptible cultivars (J.A. Dunn and Kempton 1974). The crop cultivars resistant to insect vectors of the plant viruses are likely to alter the population size, activity and probing, and feeding behaviour of vector, thereby influencing the pattern of virus spread (Kennedy 1976). Rice variety IR8 is resistant to the rice delphacid, *Sogatodes oryzicola*, in Columbia but is susceptible to the hoja blanca virus transmitted by this insect. Fields of IR8 rice remained virtually virus-free, apparently because of its resistance to the vector (P.R. Jennings and Pineda 1970). N.D. Holmes and Peterson (1957) demonstrated the depressive effect on resistant wheat of continuous rearing of the wheat stem sawfly. Infestation levels of hessian fly in the mid-western United States (where the majority of wheat is planted to resistant varieties) have dropped from more than 90% to less than 10% as the planting of resistant cultivars has increased. Resistant varieties affect cumulative reductions in pest populations. Insects infesting varieties possessing antibiosis type of resistance exhibit a high mortality, lay fewer eggs, and have slower rates of growth and, in general, smaller body size. This is also true for insects feeding on non-preferred varieties of plants, but to a lesser degree. Even normal adults from susceptible hosts, when caged on non-preferred plants, frequently lay, a lower number of eggs than they do on preferred hosts; for example, an ovipositional arrestant for boll weevil is present in certain cotton varieties (Maxwell et al. 1969).

The task of developing resistant varieties is difficult in view of the fact that the many species of parasites cover a large number of genetically different physiological races with a different degree of virulence for different kinds of crop plants or for varieties within a single species of crop. The new races are being produced continually by mutation, recombination and by some — what obscure kinds of genetic change. Some of the parasites have tremendous power of multiplication and rapid dissemination. Therefore, even well-established resistant varieties are continually endangered by new races or biotypes, produced either in the region where the varieties are grown or introduced from elsewhere by means of wind, insects, man or other agents of dissemination.

It is now clear that the best way to reduce catastrophic losses due to attacks of disease and insect pests is production of resistant cultivars of cereals and other basic food crops. However, the major pathogens/pests consist of vast and shifting populations of parasitic races. Consequently, the value of resistant varieties varies with variations in the geographical distribution and prevalence of the races to which they are either resistant or susceptible. Accordingly they may be resistant at some times and in some areas, but not in others.

2 Concepts in Disease Resistance

Disease is the result of interaction between host and pathogen. A pre-condition for the development of disease-resistant varieties is that the plant breeder is able to understand at least the mechanisms of variability in the important plant pathogen(s), including fungi, bacteria and viruses with which he is confronted in developing resistant cultivars. The role of this variation to the plant breeders, along with the meaning of various terms used in the literature is available in text books on plant pathology. However, an attempt in this chapter is made to explain these aspects in brief as a ready reference. The second, equally important, biological entity with which the breeder's concerned is the host plant. The meaning of resistance has been described by an array of terms in literature. Different kinds of mechanism cause variation in the resistance of the host plants. Of these, the type of mechanism chosen is extremely important. The second part of this chapter thus deals with types and mechanisms. Here, higher plants have been considered as the hosts, and discussion will be centred around parasites like fungi, bacteria and viruses only.

2.1 Variability in Plant Pathogens

Plant pathogens show inherited variability, which must always be considered when breeding for resistance, because new forms may evolve that can attack previously resistant cultivars. When a progeny of the pathogen exhibits a characteristic that is different from those present in the parental individual(s), it is called a *variant*. The population of genetically identical individuals produced by the variant is called a *biotype*. The biotype with certain common characteristics form a physiological race or *strain* or *pathotype*, this latter being the more correct term, although less widely used than "physiological race" to describe individuals which have a particular pathogenicity in common (R.A. Robinson 1969). *Strain* is the most common term for variants of plant pathogenic viruses. The races (or strains) of the plant pathogens are usually distinguished from each other on the basis of their pathogenicity. The *pathogenicity* is the ability of a pathogen to cause disease, while *pathogenesis* is the chain of events that lead to the development of disease in the host. The differences in pathogenicity are known on certain selected cultivars of the host called *differential* varieties.

Most changes in the characteristics of the pathogens are brought about in nature by chance and the frequency of changes favourable to the pathogen approximately equals the frequency of unfavourable ones. The probability, however, that a strain of a pathogen more virulent than the parent(s) will be produced is lower than the probability that the variant strain will be less virulent than the parent(s). A strain more virulent than the parental strain will produce disease more easily and will multiply better than the parent and will soon suppress a less virulent strain.

Virulence is the measure of degree of pathogenicity of an isolate or race of the pathogen. Some of the races may be non-virulent (avirulent). According to Vanderplank (1975b), virulence and avirulence in the pathogen are the counterparts of vertical susceptibility and resistance in the host. To most bacteriologists and virologists, a virulent strain of the pathogen causes more severe symptoms of the disease than the avirulent one. However, mycologists have used the term in a different sense, a virulent strain or physiologic race of a fungal pathogen being one which carries virulence genes which enable them to attack a particular host genotype; an avirulent race cannot attack this genotype (Russell 1978). It has recently been used to denote qualitative rather quantitative differences in pathogenicity of races.

Aggressiveness has been used to describe the capacity of a parasite to invade and grow in its host plant and to reproduce on or in it. Aggressiveness and unaggressiveness in the pathogen are the counterparts of horizontal susceptibility and resistance in the host (Vanderplank 1975b). This is also used as a measure of pathogenicity.

For disease to occur, the pathogen must be virulent. It may be virulent and strongly aggressive or virulent and unaggressive (weakly aggressive) (Vanderplank 1975).

Virulence involves gene diversity, probably largely through mutation. Aggressiveness may well involve enzyme dose (distinct from enzyme diversity) and the switching on and off of enzyme action (Vanderplank 1975).

Isolates of *Ustilago maydis* (D.C., Canada) are known to differ in the degree to which they can parasitize maize and, as no differential reaction occurs between pathogen and host, these differences must be differences in aggressiveness (Bassi and Burnett 1980). These authors also observed that aggressiveness had a low heritability and was determined by genes exhibiting mainly non-allelic interaction. High aggressiveness can be combined with low pathogenicity, as in some obligate parasites which invade the plant efficiently but cause only minimal damage to it, at least in the early stages of attack.

Pathogenicity, virulence, and aggressiveness of 282 isolates of *Septoria nodorum*, causal agent of glume blotch of wheat, were determined on eight wheat cultivars of varying resistance; cultivar reactions, measured as percentage necrosis of seedling leaves, were classified into 253 different resistance patterns. Ninety five isolates were pathogenic to all eight cultivars; 85 to seven; 54 to six; 32 to five; 11 to four; four to three; one to two cultivars. An isolate was considered pathogenic if it caused necrosis in a greenhouse test. The virulence of the isolates varied; maximum necrosis induced in the cultivars after 12 days ranged from 13 to 80%. Isolates from commercial fields were less variable in pathogenicity and more virulent than those from research plots. Aggressiveness, measured as the time to produce first necrosis on seedling leaves of the cultivar Potomac varied from 3 to 10 days after inoculation. Pathogenic and virulent types were present in differing frequencies and distributions in sub-populations defined in terms of geographic origin and year of isolation (Allingham and Jackson 1981).

2.1.1 Variability in Fungal Pathogens

The individual genotypes of a fungal pathogen may differ in many inherited characteristics, for example in morphology, physiology and pathogenicity (Person and Ebba
1975). Most of these characteristics are governed by nuclear genes (Fincham and Day
1971). Some features, however, are controlled cytoplasmically (Jinks 1966).

To the plant breeder the variability, which concerns with the development of the
new physiologic races, is of importance. A particular physiological race of a pathogen
may comprise many different genotypes, which have in common a virulence gene, but
differ from one another in morphology and physiology.

New races of fungal pathogens are produced (a) by recombination of nuclear genes
during sexual reproduction; (b) by reassortment or exchange of genetic material in
somatic cells; (c) by mutation; and (d) by extrachromosomal variation.

a) **Hybridization.** This is the combination of dissimilar gametes and incorporation into
the progeny of genetic characteristics derived from both parents. Recombination of
the genes of the two parental nuclei takes place in the zygote, and the haploid nuclei
or gametes resulting after meiosis are different both from gametes that produced the
zygote and from each other.

Thus every diploid pathogen individual is generally genetically different from any
other pathogen, even within the same species. Furthermore, the gametes produced by
such genetically different individuals will also differ in some respects from their parents
and from each other and therefore the variability of the new individual pathogens is
continued indefinitely.

A synthetic hybrid between race 7 of *Puccinia graminis* f.sp. *avenae* and *P. graminis*
f.sp. *agrostidis* was obtained by crossing the two rusts on barberry. The hybrid differed
from both parents in spore size, colour, pathogenicity and virulence. It attacked oats
and *Agrostis alba* to a limited extent (Cotter and Roberts 1963). Green (1971) hypothesized that the hybrids between *P. graminis* f.sp. *tritici* and *P. graminis* f.sp. *secalis*
resembled a more primitive form of *P. graminis* and that evolution in the stem rusts of
cereals is progressing from low virulence and wide host range to high virulence and
reduced host range.

Genetic recombination between isolates obtained from single oospore cultures of
different pathogenicity and compatibility had occurred in *Phytophthora infestans*
(Romero and Erwin 1969).

b) **Heterokaryosis.** This is the condition in which fungus hyphae or parts of hyphae
contain nuclei that are genetically different. In many fungi, particularly the asexual
ascomycetes and the fungi imperfecti, the plant body is multinucleate, at least during
the stages in the life cycle when growth is active. Furthermore, hyphal fusions in which
nuclei are exchanged between different mycelia, regardless of their sex or mating type,
are a regular occurrence in these fungi. Opportunity thus exists for genetically dif-
ferentiated nuclei to come together in the same cell to form heterokaryons, a phenomenon known as heterokaryosis. Tinline (1962) reported that the heterokaryons were
formed by hyphal anastomosis and nuclear migration in *Cochliobolus sativus*. Some
conidia, not more than 6% from heterokaryotic cultures, perpetuated the heterokaryons.

Uredospores of two biotypes of black stem-rust fungus were mixed mechanically and dusted on a compatible variety of wheat (R.R. Nelson et al. 1955). The first generation of uredospores from a compatible variety was transferred to resistant varieties, and one biotype, originating from a mixture of races 38 and 56, was found to be highly pathogenic on Khapli Emmer, a wheat previously resistant to all known North American races of stem rust.

Heterokaryotic fungi can adjust themselves to circumstances as they develop. Puhalla (1969) described a method of inducing diploid formation in *Ustilago maydis* on agar medium. He also illustrated the ease with which cell fusion and at least a transient heterokaryosis occur on agar.

c) **Parasexualism.** This is a process by which a system of genetic recombinations can occur within fungal heterokaryons. This comes about by the occasional fusion of the two genetically different haploid nuclei of the heterokaryon and the formation of diploid nucleus. Multiplication of the diploid nucleus by mitosis produces hyphae, spores and cultures containing similar diploid nuclei. During multiplication, however, crossing-over occurs in approximately 1 out of 500 mitotic divisions and results in the appearance, eventually, of genetic recombinants. This happens by separation of the diploid nuclei into their haploid components (haploidization), which occurs at the rate of about 1 in a 1000, and which results in the production of hyphae and cultures with haploid, recombinant nuclei carrying genetic material of both original components of the heterokaryon.

It is easily seen that the combination of these two processes is qualitatively exactly equivalent to the usual sexual cycle, since together they involve diploidization, recombination, and haploidization (reduction division). The only significant difference is the absence of a precise time sequence in the parasexual cycle. Crossing-over and haploidization are not two parts of a single process, following in that order, as they do in meiosis. Rather they occur independently and usually, but not necessarily, in different nuclei. Parasexuality provides for the reassortment of all these diversities both in haploid and in diploid condition, ready for natural and artificial selection (Pontecorvo 1956). Parasexuality is probably also important in generating and stabilizing variation in pathogenic fungi, particularly in those without sexual reproduction (Tinline and Mac Neill 1969; P.R. Day 1974).

Green and Kirmani (1969) indicated the possibility of somatic segregation in a uredial culture from an aecia of orange pustule of *P. graminis* f.sp. *avenae* in the greenhouse. Somatic recombination in the leaf rust of wheat caused by *P. recondita* was studied in the mixture of four races. The mixture of race 107 cm × 17 produced seven races (three of them new to India), the mixture of races 107 cm × 63 produced six races (two of them previously undescribed and two others new for India). Two other mixtures did not produce new races (S.K. Sharma and Prasad 1970). Different races of oat crown rust, *Puccinia coronata* f.sp. *avenae* were recovered from a single-spored culture of races 228 and 393. A further sub-culturing yielded non-parental virulence combinations, which could have resulted from somatic recombination (Bartos et al. 1969).

Sometimes very complex races originate, where the fungus is functionally asexual. Mixtures of races grown together on a susceptible host combine genetically to produce new races that have the sum of the parental specificities. The mechanism is obscure, but because the zoospores of *P. infestans* are uninucleate, nuclear fusion is implied. This shows that any susceptible potato variety that harbours two or more complex races may generate still higher levels of complexity (Malcolmson 1970).

The rates of mitotic recombination in *Saccharomyus cerevisae* was studied at the loci ade 3 ade 5-7, ade 6, and ade 8 by the method of increasing proportion of variants with growth. The values (per cell generation) were 3.5×10^{-4} for ade 8, 1.4×10^{-4} for ade 5-7, 4.0×10^{-5} for ade 3, and $< 2 \times 10^{-6}$ for ade 6 (Thornton and Johnston 1971).

It is generally considered that the production of new strains of rust by somatic hybridization cannot be adequately explained by hyphal anastomosis and nuclear exchange. Such a mechanism would account for the observed recombination if, as is proposed in the rust fungi, in their dikaryophase they carry multiple haploid genotypes within any single phenotype. Such a hidden variation could arise through whole-chromosome exchange between the nuclei of a dikaryotic cell (Hartley and Williams 1971).

d) Mutation. This is the change in the genetic material of a cell which is heritable to the progeny. The rate at which new variants of a pathogen are produced will depend on the mutation rate of genes at a particular locus. The mutation rate varies from gene to gene and from pathogen to pathogen (R.A. Robinson 1971; P.R. Day 1974). Mutations can be spontaneous or induced.

Flor (1956b) produced new physiological races of *Melampsora lini* by treating a dikaryon heterozygous for certain virulence genes with ultraviolet radiation. Uredospores of *M. lini*, race 1, were irradiated with ultraviolet rays, X-rays, γ-rays and fast neutrons in a quantitative study of the frequency of induced mutations at the A MaM locus (avirulence dominant to virulence on the flax variety Dakota) for pathogenicity. Mutations at this locus were identified by their ability to infect the immune variety, Dakota. The mutation frequency was proportional to dose of UV radiation and nearly proportional to dose in neutron treatments. In X-ray or γ-ray experiments, the frequency varied as the square of the dose. The mutations induced by the ionizing radiations were predominantly deletion mutations, involving loss of a chromosome segment bearing the dominant AM allele. More localized point changes are indicated by UV data. The average frequency (percent of infections) induced by fast neutrons, X-rays; and UV rays was about 2.0, 1.5, and 0.3%, respectively (Schwinghamer 1959). The genomes for virulence were induced in a flax rust culture by X-ray irradiation. The mutations are attributed to the deletion of a portion of the chromosome carrying the closely linked genes for virulence to Kota, Abyssinian, and Leona. Virulence appears to be conditioned by the absence of a dominant avirulence gene (Flor 1960b).

Mutants for virulence of race 0 of *Cladosporium fulvum* of tomato were produced with the treatment of X-rays and ultraviolet radiation (P.R. Day 1957). He further cautioned that unintentional treatment of pathogens may occur with the wholesale treatment of plant material with radiation. Baker and Teo (1966) induced mutants affecting uredospore colour in *P. graminis* f.sp. *avenae* by using ethylmethane sulphonate. The colour mutants were recessive and EMS was an extremely efficient mutagenic agent.

Variation in virulence of blast fungus of rice *(P. oryzae)* was induced by X-rays (Kwon et al. 1974). The mutation frequency was directly proportional to the radiation doses. The X-ray increased the variability for pathogenicity.

Eleven asexual spontaneous variants of *M. lini* were observed by Flor (1960b). Green (1964) isolated mutants for colour and virulence from the teliospores of *P. graminis*. A mutation in *P. recondita* f.sp. *tritici* to virulence on transfer, Chinese spring × *Aegilops umbellulata* was observed (Samborski 1963). A mutation for virulence in race 107 of brown rust of wheat, *P. recondita*, was observed after 400 uredinial generations in glasshouse (S.K. Sharma and Mishra 1965). Mutations for virulence in *P. coronata* f.sp. *avenae* were studied in the clones of races 202, 203 and 290 (Zimmer et al. 1963). The races were cultured continuously in isolation on a susceptible oat variety. Inoculation of 202 and 290 on five highly resistant host varieties in the 5th through 9th uredial generations after clonal establishment led to the production of rare non-parental-type pustules. Race 203 did not produce non-parental pustules. The pathogenicity characteristics of all variants and their recurring production suggested mutations as the operative mechanism. The mutation rate of race 202 for virulence on Ascencao was roughly estimated in 1 out of 2200 infections; for virulence on Ukraine, in 1 in 6450; and for race 290 for virulence on Ukraine in 1 in 7900.

Franzone and Favret (1982) induced mutations for higher and lower virulence in *P. recondita* f.sp. *tritici* by ethylmethane sulphonate. The mutation rate for higher virulence was 50 times higher than the mutation rate for lower virulence.

Single spore isolates of *P. coronata* race 228, avirulent on oat variety Saia were increased and screened on Saia. Eighty seven infections were obtained from a total of nearly 40,000 infections. This variation could result from a high mutation rate (Bartos et al. 1969).

According to Van der Plank (1963), a mutant of a pathogen with virulence towards a resistance gene of a previously resistant host variety will quickly become established in the pathogen population, if that variety is widely grown. In the absence of the variety, however, the fate of a new physiological race will be determined partly by its ability to compete successfully with other genotypes of the pathogen, i.e. its fitness to survive (Van der Plank 1968).

e) Cytoplasmic Adaptation. Adaptation is the acquisition by a pathogen of the ability to carry out a physiological process which it could not carry out before, or at least not effectively. Although pathogens may acquire such abilities by means of mutation in their genetic material in the nucleus, there are several examples of cytoplasmic inheritance of important characteristics, such as growth rate and virulence (Jinks 1966). Virulence of *P. graminis* f.sp. *avenae*, the stem rust fungus to oat varieties carrying resistance gene E, is maternally inherited (Green and McKenzie 1967) and may be controlled by single plasmagene (T. Johnson et al. 1967). Aneuploidy or cytoplasmic inheritance might be responsible for the variation in virulent isolates of race 229 of *P. coronata* f.sp. *avenae*, when sub-cultured on oat variety Saia (Bartos et al. 1969). Colonies derived from white sporidia of *U. maydis* became brown when in contact with sporidia capable of forming a browning agent. The factor responsible for the production of browning may be the cytoplasmic factor (Anagnostakis 1971).

2.1.2 Variability in Bacteria

Four different well-defined modes of transmission of genetic material have been discovered in bacteria, which lead to variation in pathogenicity.

a) **Conjugation**. Reproduction of bacteria is usually asexual and involves the division of one cell into two cells. Two compatible bacteria come in contact with each other and a small portion of the genetic material (DNA) of one bacterium is transferred to the genetic material of the other. In this way the genetic make-up of both cells is altered. Binary fission of these cells produces daughter cells of different genetic characters and these develop into new races by asexual reproduction. In many bacteria there are DNA fragments in the cytoplasm in addition to the large circular chromosome. Some of these fragments may be incorporated into the main chromosome during cell division, forming new bacterial genotypes.

b) **Transformation**. Bacterial cells are transformed genetically by absorbing and incorporating in their own cells genetic material secreted by, or released during rupture of, other compatible bacteria. The incorporated genetic material changes the properties of the receptor bacterium by the number of genes added to it, and a new strain results. Harris-Warrich and Lederberg (1978) reported interspecific transformation in *Bacillus* mechanism of heterologous intergenote transformation.

c) **Transduction**. A bacterial virus (bacteriophage or phage) transfers genetic material from the bacterium in which the phage was produced to the bacterium it infects next. If the second bacterium is not killed by the phage, the additional genetic substance is incorporated into the existing genetic material of the bacterium and is thereafter transmitted to its descendants, which constitute a new strain.

d) **Mutations**. The most important source of variation in bacteria are mutations. In a culture of *Escherichia coli* there is, on average, one mutant cell in every 2000 cells. Bacteria can reproduce very quickly under optimal conditions and the multiplication rate of mutant types would be very rapid if they were strongly favoured by selection pressure, as, for example, are plants with race-specific resistance. Dahlbeck and Stahl (1979) reported mutation for change of race in culture of *Xanthomonas vesicatoria*. Mutation occurred randomly in nutrient broth culture at an apparent rate of 4×10^{-4} mutation/cell/division.

2.1.3 Variability in Viruses

Viruses are not considered living organisms. However, the molecular structure of viruses has the capacity of transferring its characters to its duplicates produced in the host cells.

a) **Hybridization**. This is one of the methods by which new virus strains are formed. If two strains of virus are inoculated into the same host plant, one or more new virus strains may be recovered with properties (virulence, symptomatology etc.) different from those of either of the original strains introduced into the host. These new strains are probably hybrids (RNA or DNA recombinants). Albersio et al. (1975) reported variability in squash mosaic virus (SMV) by hybridization between two strains of virus. They had crossed strain I H and II A of SMV in pumpkin *(Cucurbita pepo)* and cantaloupe *(Cucumis melo)* plants and observed the interaction between them.

b) **Mutations**. The evolution of new strains of viruses may also be due to mutation. These may be a heritable change in the genetic material (RNA or DNA). The production of mutants differing in virulence has also been reported in several viruses, especially TMV, although they seem to vary mostly in the type of symptom and severity of disease they produce rather than in their ability to infect different host plant varieties.

2.2 Identification of Physiological Races

As early as 1894, Erickson in Sweden demonstrated that the micro-organism stemrust, *Puccinia graminis*, varies in pathogenicity. He noted that black stem rust taken from wheat did not infect oats, rye, and some other grass species. He next established that, in collections taken from various hosts, each had a characteristic host range, attacking certain species but not others. On this evidence he defined several sub-species of *P. graminis* based on differing physiological properties as expressed in specific pathogenic properties on different grass hosts. Barrus (1911) described physiological races on the basis of differing pathogenicity on different varieties of the same host species. On the basis of differential reactions to bean varieties *(Phaseolus vulgaris)*, he distinguished two races, denoted α and β, of bean anthracnose *(Colletotrichum lindemuthianum)*. Shortly thereafter Stakman (1914) showed that Erickson's sub-species were not pathogenically homogeneous, but, like the bean anthracnose organism, included physiological races varying in their pathogenicity on different varieties within a single cereal species. Evidence rapidly accumulated, largely through the efforts of Stakman and his collaborators, that each of Erickson's sub-species was comprised of many physiological races.

Prasada et al. (1964) reported the occurrence of a virulent biotype of race 162 of *P. recondita*. They reported differences in pathogenicity on Henry and Thew varieties of wheat. For biotype 162 A Thew was susceptible, but was resistant to 162 B.

Physiological races can be identified by their reaction on a group of varieties known as differential hosts (Young and Browder 1965). To show the heterogeneicity of *P. graminis* f.sp. *tritici* races, supplementary differential sets of NA 61 and NA 65 American wheat varieties and single pustule cultures of races 77, 130 and 184 were used. Different races were shown to have in their composition NA65-23 sub-race present in all the three races tested (Florica 1971).

Originally the differential hosts were chosen with the knowledge that they differed in resistance, but with no information on the genetic basis of the resistance. If only two possible infection reactions, resistance and susceptibility, are recognizable, a set of n differential hosts (tester genotypes), whose reaction to infection is governed by a

Table 2.1. Dependence of race identification on number of host testers (only two infection reactions are assumed: 0 = resistant, + = immune). (After Williams 1964)

No. of testers	Host tester	Physiological races																No. of races
		1	2	3	4	5	6	7	8	9	10	11	12	13	14	15	16	
1	A	+	0															2
2	A	+	+	0	0													4
	B	+	0	+	0													
3	A	+	+	+	+	0	0	0	0									8
	B	+	+	0	0	+	+	0	0									
	C	+	0	+	0	+	0	+	0									
4	A	+	+	+	+	+	+	+	+	0	0	0	0	0	0	0	0	16
	B	+	+	+	+	0	0	0	0	+	+	+	+	0	0	0	0	
	C	+	+	0	0	+	+	0	0	+	+	0	0	+	+	0	0	
	D	+	0	+	0	+	0	+	0	+	0	+	0	+	0	+	0	

single gene pair, can differentiate 2^n races, where n is the number of testers (Table 2.1). A simple host-parasite relationship which allowed the host reaction to be classified into two easily defined reaction grades has been used in both *P. infestans*, the late blight fungus of potato, and in *M. lini*, the flax rust fungus. If five levels of pathological reaction can be recognized on each host, n differential hosts are potentially capable of differentiating 5^n races. Loegering and Browder (1971) suggested a system of nomenclature of physiological races of *P. recondita*. Different wheat varieties may be assigned sequential numbers in avirulence/virulence formulae. In routine surveys, race nomenclature will be obtained by giving codes to formulae based on any set of differential varieties. Nomenclature of important races may be based on type culture preserved in liquid N. W. Black (1960) reported that six testers were available for racial identification within potato blight, therefore, 64 races could be identified. Flor (1956) reported a large number of testers for racial identification within flax rust and 179 races of *M. lini* were characterized.

Greater precision has been achieved by using hosts having known genes for resistance. A system used for *P. infestans* may be ideal, where a physiological race is designated by the virulence genes it carries or by the resistance genes it can overcome, but it cannot be applied to the many host-pathogen relationships in which the gene-for-gene hypothesis may not hold. For these pathogens the differential host remains the basis of identification (Habgood 1970).

Information from China, Indonesia, Japan, Korea, the Philippines and other countries indicates that there is physiological specialization of *Xanthomonas campestris* pv. *oryae*, the rice *(Oryza sativa)* bacterial blight pathogen, on a set of differential varieties. Based on the rice-bacterium combination, the rice varieties showed strong interaction with bacterial isolates in some countries, but weak interaction in others (Mew et al. 1982). Eamchit and Mew (1982) distinguished the isolates of *X. campestris* pv. *oryzae* from Thailand and the Philippines into groups of strain virulence on rice cultivars with different genes for resistance to the pathogen. All groups were avirulent on cultivar DV 85 (recessive gene xa-5 and dominant gene Xa-7). Group 0 was virulent on all cultivars including IR 8 and RD 9 that carry no genes for resistance to the pathogen. The Thai group I was comparable to the Philippine group I and was virulent only on IR 8 and RD 9. The Philippine group II was similar in virulence to the Thai group III, but differed from it in that it was virulent on both RD 7 (genes not analyzed) and IR 20 (dominant gene Xa-4), the Philippine group II was virulent on RD-7 but not on IR20. Only the Philippine group III isolates were virulent to IR 1545–339 that has the recessive gene xa-5 for resistance to the pathogen. Group II and III were virulent on IR 1695 (dominant gene Xa-6) and PI 231129 (recessive gene xa-8): both cultivars have adult plant resistance to *X. campestris* pv. *oryzae*. It is suggested that pathogen specialization has occurred and that the specificity of infection of these differential cultivars can be distinguished in the vegetative stage by lesion length or area.

Browder et al. (1980) observed that race, as applied to studies of host-parasite specificity, is a group of individuals in a parasite population with pathogenicity characters in common. Therefore, the term race may be used for that meaning only. No term is needed to convey the concept taxon with specified pathogenicity, because the formal taxon is not needed. They further observed that studies of pathogenic specialization in systems in which genetics are not known would be more useful if their objectives were to elucidate genetic relationships rather than merely to describe and name parasite variation for pathogenicity to host cultivars.

2.3 Genetic Nature of Physiological Specialization

Studies on the nature of the genetic differences between physiological races have been undertaken in many species of pathogenic fungi including several rusts (*Puccinia* spp. and *Melampsora lini*), the smuts (*Ustilago* spp.), apple scab *(Venturia inaequalis)* and many others.

The first study on the genetics of pathogenicity in fungi is usually ascribed to Goldschmidt (1928), who crossed two races of anther smut, *Ustilago violacea*. The hybrid between a race attacking only *Silene saxifraga* and one attacking *Melandrium album* was shown to be capable of attacking both host species. The genetic nature of physiological races of *P. graminis* f.sp. *tritici* was investigated by Newton et al. (1930). Eight races were self-fertilized and all except one segregated for infection reaction, showing that the majority of the races studied were heterokaryotic for the infection reaction. Certain of the races proved to be homokaryotic for infection reaction on some tester varieties and heterokaryotic on others. In other words, selfing failed to give rise to segregation of reaction type on some testers, while on others segregation was observed. The presence or absence of segregation, depending on the tester used, indicated that more than one gene was responsible for the control of infection of these races. Some of the non-pathogenic races give rise on selfing to pathogenic segregates, establishing that pathogenicity on some tester varieties was recessive to non-pathogenicity. For example, a race that failed to infect the variety Kanred produced, on selfing, a race to which Kanred was fully susceptible.

2.4 The Importance of Parasitic Variability to the Plant Breeder

The ability of many plant pathogens to become adapted to previously resistant varieties has been one of the main problems encountered by plant breeders. This is because of the variant races or isolates of pathogens which are able to overcome the resistance and become adapted on the cultivated resistant varieties. Therefore, the resistance of such varieties breaks down; for example, wheat varieties Kanred, Kota and Ceres showed good resistance to stem rust when they were first cultivated in the USA, but they succumbed to new physiological races of *Puccinia graminis*; as a result, millions of acres of wheat in the USA were devastated by stem rust. Similarly, the resistant varieties of wheat and potato to yellow rust *(P. striiformis)* and late blight *(P. infestans)*, respectively, were attacked by virulent physiological races of these pathogens.

Strains of viruses and bacteria that are able to attack resistant varieties have generally been less of a problem to plant breeders than physiological races of fungal pathogens.

2.5 Resistance in the Host

The extent to which a plant prevents the entry or subsequent growth of the pathogen within its tissues or the extent to which a plant is damaged by a pathogen is used to measure the resistance or susceptibility.

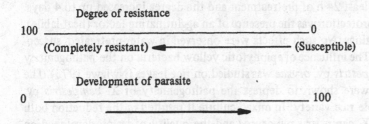

Fig. 2.1. Definition of resistance. (Eenink 1977)

Resistance may be qualified by such words as high, intermediate or low because there may be gradations between extreme resistance and extreme susceptibility. The resistance may be conditioned by a number of external and internal factors operating to reduce the chance and degree of infection. The infective agent is dependent on finding a favourable physical and chemical environment in its host from the time of first contact until the completion of its life cycle, while the host, if it is to resist attack, must possess some physiological or morphological characteristics that will inhibit or destroy the parasite at some stage in its development.

2.5.1 Mechanisms of Resistance

a) **Induced Resistance**. When adequate dosages of uredospores of *Uromyces phaseoli* were placed on sunflower leaves before inoculation of the leaves with *Puccinia helianthi* or along with the inoculum, the sunflower leaves were protected from infection with *P. helianthi*. Similarly, spores of *P. helianthi* protected bean leaves from infection with *U. phaseoli*. To give 50% protection of sunflower from *P. helianthi* under conditions of these tests required about 1.3 mg of uredospores of *U. phaseoli*/square diameter of sunflower leaf surface. Yarwood (1956) called it cross-protection. Induction of resistant response (hypersensitivity) by inoculation with *Drechslera (Helminthosporium) carbonum, Alternaria* sp. (the non-pathogens) or varietal non-pathogenic races of *Colletotrichum lindemuthianum* renders the inoculated tissues resistant to varietal pathogenic races of *C. lindemuthianum* (Rahe et al. 1969).

Flax rust infection was significantly reduced as a result of prior inoculation with an avirulent race of the pathogen. The induced resistance was a localized effect expressed as reduction in the size, number and rate of development of uredia of the virulent race. An interval of 4 h between successive inoculations was enough to induce resistance, but the reaction intensified as the interval between inoculations was increased to 24 h.

The induced resistance was effective for 7 days after inoculation with the avirulent race. The diffusable substances from the pathogen may be responsible for triggering the induced resistance reaction (Littlefield 1969).

Induced resistance to bacterial pathogens is also reported. Main (1968) noted that susceptible bottom cuttings pretreated by stem uptake inoculation with suspensions of avirulent mutants (B_1 and B_2) of *Pseudomonas solanacearum* K60 developed protection against wilting when subsequently challenged with the virulent bacteria. Onset of protection required at least 24 h of pretreatment and the degree increased up to 4 days. Associated with this protection was the presence of an agglutinating factor (heat-labile), isolated from the cuttings. No such effects were observed in water-pretreated, susceptible control cuttings. The influence of epiphytotic yellow bacteria on the pathogenicity of *Xanthomonas campestris* pv. *oryzae* was studied on rice leaves (Nwigwe 1973). The epiphytotic bacteria were shown to depress the pathogenicity of *X. campestris* pv. *oryzae* on a susceptible rice variety. In mixed culture it resulted in the reduction both of the infectivity of *X. campestris* pv. *oryzae* and the quality of lesion developed on the host plant.

b) Escape. The coincidence of critical phases of development in the host and its parasite frequently governs the magnitude of disease problems. Certain varieties of crop plants which rapidly undergo development and maturation may complete their life cycle before maximal infection occurs; for example, early maturing varieties of wheat may escape serious damage by rusts in areas where the disease develops late in the growing season. Morphological and physiological characters which allow consistent disease escape are, however, comparatively rare in plants where they can be utilized in breeding, as in the case of the closed flower varieties of barley which have general escape mechanisms to infection by *Chlàmydospores* of loose smut *(Ustilago nuda)* of barley (Macor 1960). Varieties such as Maythorpe, Proctor and Spratt Archer possess a high degree of resistance in the field, and are completely susceptible when inoculated with spore suspensions. Therefore, the field resistance in this case is associated with some form of escape mechanism. Studies on flowering revealed that in the field-resistant barley varieties, 95–98% of the flowers remained completely closed, thus preventing infection through the stigma and ovary. By contrast, the field susceptible varieties, such as Freja, have a high proportion (up to 25%) of fully opened flowers at maturity which allow direct infection through the ovaries and stigmas. Certain corn hybrids produce ears that turn outward and hang down, and thus prevent the occurrence of favourable micro-environmental conditions necessary for certain ear-rotting organisms. Varieties of raspberries susceptible to virus causing raspberry mosaic frequently escape the disease because the plant is a non-preferred host of the insect vector. The term klenducity is used commonly to characterize the type of disease escape accomplished by the avoidance of the insect vector (R.R. Nelson 1973).

Several leaf characters of rice cultivars like rolled, narrow, dark green and slow senescence were associated with resistance to bacterial leaf blight (*X. campestris* pv. *oryzae*) resistance (C.B. Singh and Y.P. Rao 1971).

c) Tolerance. A plant which is attacked by a pathogen to the same degree as other plants, but which suffers less damage in terms of yield or quality as a result of the attack, is said to be tolerant (R.A. Robinson 1969). Tolerance may be defined as the inherent or acquired capacity to endure disease (R.R. Nelson 1973) and to give satisfactory returns to the grower (M.V. Rao 1968). Tolerant plants are susceptible to infection, to spread and colonization of the parasite, but they exhibit tolerance of the parasite by resisting the impact of disease. The nature of tolerance is not clearly known. It may be in the form of compensating growth as in corn roots infected by root-rotting pathogens, or may be in the form of bearing heavy infection and yet giving almost normal yields, as in wheat attacked by rust (M.V. Rao 1968).

The concept of tolerance cannot be applied to diseases that affect the end-product of the host. There can be no tolerance in the cereals to smut pathogens that attack kernels. Cultivars of alfalfa can exhibit no tolerance to foliage-attacking pathogens if the crop is grown and harvested for hay. The same rationale applies to the various diseases of fruits, vegetables etc. Van der Plank (1963), however, considers that tolerance is not a desirable character because it does not control either initial inoculum (X_0), infection rate (r) or infection time. However, it may be useful as a floor in combination with other resistance.

Resistance mechanisms that rely entirely on host tolerance, rather than on resistance to infection, have the serious disadvantage that they serve as potentially dangerous reservoirs of inoculum and, although they themselves may be satisfactory, they increase damage on other varieties.

Russell (1978) observed that for viral diseases there are at least three different kinds of tolerance mechanism:

1. The virus is able to multiply but symptoms do not appear. In virus yellows of sugarbeet in spite of the absence of visible symptoms on the leaves, the plants harbouring virus suffered significant losses of root or sugar yield (Russell 1964). The cultivars in which infection damages the host but does not produce symptoms are called symptomless carriers.

2. The plants which develop disease symptoms which are as severe as those in other plants but which suffer less damage, referred to as disease-tolerant. Many breeding lines of sugarbeet show severe yellowing symptoms on the leaves when infected with virus yellows, but nevertheless yield large roots with a high sugar content.

3. The virus-infected plants that do not show severe disease symptoms and are less damaged by infection than other plants are true-tolerant, for example, infected barley plants that are tolerant to barley yellow dwarf virus do not show pronounced yellowing or stunting of the leaves and give an acceptable yield of grain in spite of the barley being infected.

Virus tolerance is simply inherited (dominant/recessive) in many virus-host combinations, for example, barley yellow dwarf tolerance is controlled by an incompletely dominant gene and a recessive gene; bean yellow mosaic virus by two or three major genes in *Phaseolus* (Bagget 1956) and by a single dominant gene in red clover (Diachun and Hensen 1959). However, for yellow and curly top virus in sugarbeet continuous variation for tolerance was observed (Russell 1978).

Simons (1969, 1972a) studied tolerance to crown rust in oats and concluded that the trait is quantitatively inherited. Broad sense heritability estimates of tolerance to *Verticillium* wilt *(Verticillium albo-atrum)* under severe field exposure in American

Upland cotton *(Gossypium hirsutum)* were compared using F_2 and F_3 generations of tolerant X susceptible crosses. Heritability estimates varied from 0 to 0.833, depending upon exposure level, generation tested and type of parental tolerant lines involved (C.L. Roberts and Staten 1972). Tolerance to white mould *(Sclerotinia sclerotiorum)* in ExRico 23, a cultivar of white bean *(Phaseolus vulgaris)*, was studied by Tu and Beversdorf (1982). In a field with severe white mould infection, ExRico 23 had consistently lower disease incidence and a slower rate of disease spread than other recommended entries of beans.

d) Immunity. Immunity means exempt (100%) freedom from infection, i.e. immunity of a plant to infection by a given disease is an absolute quality. It denotes that the pathogen cannot establish parasitic relationship with the host even under the most favourable conditions. An ultra-structure comparison of the susceptible and immune reactions of *Vigna unguiculata* leaves infected with *Uromyces phaseoli* var. *vignae* showed that signs of incompatibility were detectable in the immune variety during the early growth stages of haustorial formation when a deposit of callose-containing material was formed on the host cell wall around the point of entry of the haustorium. No such reaction was observed in the susceptible variety (Heath and Heath 1971).

e) Hypersensitivity. The hypersensitive reaction is operative in host plants infected with bacteria as well as fungal and viral pathogens (Klement and Goodman 1967). It is an extreme degree of susceptibility in which rapid death of cells in the vicinity of the invading pathogen (the infection court) occurs, so that the progress of the pathogen is halted although it does not necessarily die immediately. Hypersensitivity thus confers high resistance and is sometimes used in the sense of immunity. In tobacco plants, for instance, the hypersensitive reaction to tobacco mosaic virus (TMV) provides a high degree of resistance by confining and killing the virus within a primary necrotic lesion which forms following entry. Thus the reservoir of infective virus is reduced and the spread of the disease is effectively controlled. Hypersensitivity to TMV was not known among plants of *Nicotiana tabacum* ($2n = 24$), but was found within *N. glutinosa* ($2n = 12$) and the dominant gene responsible for the reaction was subsequently transferred to *N. tabacum* through the use of artificially produced amphidiploid, *N. digluta* (F.O. Holmes 1938). As a result of back-crossing the hybrids between *digluta* and *tabacum* to commercial varieties of tobacco, several acceptable commercial lines of tobacco possessing hypersensitivity to all strains of the virus have been produced and released for production (Valleau 1952).

Klement and Goodman (1967) noted that: (1) the hypersensitive reaction operates only in incompatible host-parasite combinations; (2) at the initial stage of the infection process there is no appreciable difference between the rates of multiplication of the pathogen in susceptible and resistant plants; (3) hypersensitive necrosis appears earlier in resistant plants than typical symptoms in susceptible; (4) the development of the hypersensitive reaction in plants is associated with the loss of cell turgor, which reflects a loss in cell membrane permeability; and (5) the hypersensitive reaction can be mimicked by numerous synthetic sulphydryl (SH)-containing compounds.

Hypersensitive cells produce chemical compounds, including phytoalexins, many of which are phenolic; they are not only fungitoxic but also autotoxic, so that, in effect, the cells kill themselves (Deverall 1976). A pathogen may die as a result of two distinct mechanisms, either directly by the action of the host plant toxins or indirectly because the host cells die and are unable to support an obligate pathogen. It is also suggested that hypersensitivity is the result, rather than the cause of resistance (Ogle and Brown 1971, Kiraly et al. 1972, Mayama et al. 1975). On the other hand, Bracker and Little-field (1973) and Maclean et al. (1974) considered that hypersensitivity can directly cause incompatibility.

Yamamoto and Matsuo (1976) suggested that interactions between the DNA of specific genotypes of host and pathogen may result in a hypersensitive reaction which would be a cause rather than a consequence of resistance.

Although the precise nature of hypersensitivity is not understood, this kind of resistance has been widely used by plant breeders because it is easy to exploit in breed-ing programmes. It is controlled by a few major dominant genes, and resistant plants can easily be distinguished from susceptible plants. Resistance based on hypersensitivity has given very effective control in barley for powdery mildew and in potato for late blight. However, varieties with R genes for hypersensitivity to the blight fungus, *P. in-festans* are susceptible to isolates of the pathogen that do not trigger off a hyper-sensitive response in such varieties. Similarly, barley varieties with hypersensitive resistance to some isolates of *Erysiphe graminis* have been severely attacked by other isolates (Russell 1978).

A single dominant gene (Bs3) is reported in the PI 271322 line of pepper *(Capsicum annuum)* that confers hypersensitivity against pepper strain race 1 of bacterial spot, *X. campestris* cv. *vesicatoria* (B.S. Kim and Hartmann 1985).

2.5.2 Types of Resistance

a) **Resistance Based on Mode of Inheritance.** Monogenic resistance is governed by one gene and polygenic resistance is governed by many genes. The gene in monogenic resistance can usually be studied in detail, identified individually by letters or numbers or both. Thus the R_1 gives resistance to potato blight; the Sr_6 gives resistance to wheat stem rust; and B gives resistance in oats to stem rust. The sort of resistance the gene gives, whether dominant or recessive, can be determined. Monogenic resistance is clear-ly defined, and in segregating populations the plants can be divided by their resistance into clear and natural groups. The gene can be located on a particular chromosome; the gene Sr_6 is on chromosome XX of wheat (Peterson and Campbell 1953).

The polygenic resistance is conferred by too many genes to be counted and identified individually. The plants in segregating populations vary practically continuously in resistance, without falling into clearly defined groups.

The division of resistance into monogenic and polygenic resistance makes no room for intermediate sorts, for example, digenic or trigenic resistance, which are neither monogenic nor polygenic. The proper opposite of poly-meaning many, is oligo-mean-ing few; and therefore, in terms of number of genes, one should speak of oligogenic rather than monogenic resistance. Resistance specified by genes of major effect most frequently acts singly (Person and Sidhu 1971); perhaps rarely due to several inter-acting genes such as complementary genes for resistance.

Monogenic or oligogenic resistance is also called specific resistance. It has short-lived usefulness, although there are notable exceptions (R.M. Caldwell 1968). It has generally been associated with host cell hypersensitivity, resulting in cell death or chlorosis and impairment. It has been designated by a confusing array of names, including physiological resistance (H.K. Hayes and Immer 1942), specific resistance (Stakman and Christensen 1960), racial resistance (Niederhauser et al. 1954), monogenic perpendicular and hypersensitivity etc.

Polygenic resistance has also been designated by many names, including partial resistance (Niederhauser et al. 1954), field resistance, general resistance, generalized resistance, tolerance etc.

Resistance is also defined in qualitative and quantitative terms. "Qualitative" is used to indicate whether a cultivar is resistant. The type of lesion, not the number, is considered in determining qualitative resistance. Quantitative resistance takes into account the extent of disease or the number of lesions (Ou 1979). Niederhauser and Cervantes (1956) reported that field resistance to *P. infestans* in the same clones of potato is independent of the hypersensitive or major gene resistance. The rice lines with fewer lesions were resistant to more races and isolates of *Pyricularia oryzae* (Ou et al. 1975). Thus, the quantitative difference in resistance (lesion number) observed in the field was an indicator of the qualitative (race reaction) resistance in rice. This quantitative nature of resistance resembled what is commonly referred to as horizontal or general resistance, but apparently is conditioned by the presence of several to many vertical or race-specific resistance genes (Ahn and Ou 1982).

b) Resistance Based on Effect of Genes. The terms major and minor genes are sometimes used instead of oligogenic and polygenic when referring to gene resistance. However, all oligogenes are not necessarily major genes in the sense of being important genes. On the other hand, all polygenes are not minor genes in the sense of being relatively unimportant.

c) Resistance Based on Growth Stages of the Host Plant. Resistance of wheat to leaf rust, *Puccinia recondita* sp. *tritici*, may be expressed at the "first leaf stage", is generally termed "seedling resistance", and is usually effective at all stages of plant development (Bartos et al. 1969). The term "post-seedling resistance" or "adult plant resistance" is used in the sense that a cultivar is susceptible at first leaf stage to one or more races, but at a later stage of development it becomes resistant to the same race or races (R.G. Anderson 1966).

Seedling plant reactions are used for the identification of rust races, and the genetic behaviour of many genes for seedling resistance has been thoroughly investigated and is governed by a dominant gene (Kerber and Dyck 1969). Genetic analysis of adult plant resistance is often complicated by the presence of seedling resistance gene(s), which are effective against the strains of rust used in the study. Where no seedling resistance gene(s) are involved, adult plant resistance may result from a single major gene. Nevertheless, the action of this gene may be influenced by modifying genes, resulting in a range of reaction types in segregating plant populations, which makes interpretation difficult (Dyck et al. 1966).

Resistance of adult leaf rust derived from *Aegilops squarrosa* was conferred by a single, partially dominant gene that is inherited independently of Lr12 and Lr13, two previously identified genes for adult plant leaf rust resistance. Although monogenic inheritance was observed, this gene must be influenced by the genetic background, since its level of resistance was somewhat reduced during successive backcrosses to Thatcher (Dyck and Kerber 1970).

Since the 1920's, some varieties of wheat have been known to show a striking difference between their reaction to stem rust at different stages of growth (Goulden et al. 1928). These authors also reported that seedling and adult plant resistance were independently inherited.

It was shown by Knott (1968) that wheat varieties Hope and H-44 carry two dominant genes for resistance to race 56 of stem rust and one recessive gene for resistance to race 15-1 L (Can.). One gene, designated Sr 1, conditioned seedling resistance to race 56 and is on chromosome 2B. A second gene, designated Sr 2, conditions adult plant resistance to race 56, but by itself has no detectable effect on seedling resistance. However, it acts as modifier of Sr 1, and both Sr 1 and Sr 2 must be present to provide full resistance at either the seedling or adult plant stage. Resistance to 15B-1L is controlled by a single recessive gene. Knott (1971b) found that the gene Sr 1, which conditions resistance to race 56, was found to be either very closely linked or more probably allelic to Sr 9. It was proposed to redesignate it Sr 9 d. The gene Sr 2, which conditions adult plant resistance to race 56, appeared to be on chromosome 3B. The recessive gene conditioning resistance to race 15B-1L was identified as Sr 17, which was on chromosome 7B.

Genetics of resistance of stem rust in wheat to Indian races was reported by Nathawat et al. (1979). The wheat cultivars were E 4849, E 5533, E 5550, Raj 848, Timgalen, HD 208, Safed Lerma, Tr 373, and races of stem rust were 21 and 40 (*P. graminis* f.sp. *tritici*). Seedling resistance in six of these cultivars was controlled by previously identified genes; all eight cultivars have additional (one or two) genes which seems to be new. The results indicate that the cultivars have the following genes: E 4849 − Sr 8, Sr 11, Sr 4849 (new gene); E 5533 − Sr 11, Sr 5533 (new gene); E 5550 − Sr 8, Sr 11, Sr 5550 (new gene); Raj 848 − Sr 11, Sr 848 (new gene); Timgalin − Sr 5, Sr 8, Sr Tg (new gene); HD 2028 − Sr 11, Sr 2028-1 (new gene), Sr 2028-2 (new gene); Safed Lerma − Sr S1 (new gene); Tr 373 − Sr 373-1 (new gene); Sr 373-2 (new gene).

Greenhouse evaluation of adult plant resistance in the wheat cultivars Hope, Hopps and H-44 was conducted by Sunderwirth and Roelfs (1980). The inoculation was done at first node, second node, boot and anthesis. The resistance to stem rust conditioned by Sr 2 gene was non-race-specific to the races 15-TLM, 15-TNM and 151 QSH and was characterized by reductions in number, size and location of uredia. The resistance was best expressed after anthesis.

Twenty varieties of wheat possessing known Yr genes and their different combinations were tested against 13 races of stripe rust (*P. striiformis* West) both in seedling and adult plant stage. The results indicated that genes Yr 2, Yr 3a, Yr 4b, Yr 5 and Yr P.I.178383 were effective both in seedling and adult plant stages. Varieties Chinese 166 (Yr 1) and Peko (Yr 6) appeared to have additional gene(s) effective in post seedling stage, besides the known Yr genes. Reactions of 32 genetic stocks and commercial varieties indicated that a number of genetic stocks exhibiting race-specific and non-race-specific type of resistances are available (Upadhyay and Kumar 1979). These can be profitably used in stripe rust resistance breeding programmes.

Genetic analysis of adult plant resistance to powdery mildew in pea *(Pisum sativum)* was conducted by Kumar and R.B. Singh (1981). Resistance to powdery mildew *(Erysiphe polygoni)* was controlled by duplicate recessive genes (er_1 and er_2).

Six spring barley varieties *(Hordeum vulgare)* with different degrees of resistance or susceptibility to *E. graminis* f.sp. *hordei* were compared under controlled conditions on the basis of infection types and infection rates (Hwang and Heitefuss 1982). Resistance of the cultivar Villa and Asse, which were susceptible at the seedling stage but resistant at later growth stages, was race-specific at the seedling stage. As plants or leaves of all tested cultivars, both susceptible and resistant, became older, their degree of resistance increased gradually. This increase in host resistance was marked by an apparent transition in infection type and reduced colony production, depending upon the developmental stage of plants. Differences in colony production of powdery mildew fungus on the different cultivars were detected on the seedlings, and these differences remained consistent at all later growth stages of the plants. The reduction of colony production was related to the gradual change of infection type during plant development. The cultivar Asse developed fewer colonies per unit of leaf area than the cultivars Mari S, Peruvian and Villa irrespective of race inoculum density or plant development.

An accession of *Avena sterilis*, Canadian *Avena* (CAV) 1387, which originated from Israel, was susceptible in the seedling stage but resistant in the adult plant stage to six races of crown rust, *P. coronata*. The adult plant, resistant in CAV 1387, was conferred by a single partially dominant gene designated Pc–69 (Harder et al. 1984).

Inheritance of mature plant resistance to the southern-corn leaf blight, *Drechslera (Helminthosporium) maydis* race 0 in maize, *Zea mays* was studied by D.L. Thompson and Bergquist (1984). They reported it to be independent of seedling reaction, which was described as due to relatively few loci with additive effects.

d) Resistance in Epidemiological Terms. When a variety is more resistant to some races of pathogen than to others, the resistance is called "vertical". Van der Plank (1963) illustrated this in potato varieties with R_1 gene for resistance to potato late blight *(P. infestans)*. Vertical resistance implies a differential interaction between varieties of the host and races of the pathogen. When the resistance is evenly spread against all races of the pathogen it is called "horizontal". In horizontal resistance there is no differential interaction. Van der Plank (1963) also illustrated horizontal resistance (HR) to blight of the foliage of the two varieties without R genes. The HR was considerable in Capella and small in Katahdin.

R.R. Nelson (1973, 1978) has argued that to accept this interpretation of HR leads to an empty conclusion, since there appears to be no evidence for the existance of such resistance to any plant pathogen. He therefore, redefined HR in an epidemiological context (R.R. Nelson 1978) as resistance that reduces the apparent infection rate, which, however, can also be due to partial vertical resistance, and other forms of resistance.

Ahn and Ou (1982) studied the epidemiological implications of the spectrum of resistance of rice blast. The infection efficiency (IE) and the infection rate (IR) of the spore population of *Pyricularia oryzae* was lower if the percentage of races to which each cultivar was resistant was high. It appears that cultivars with resistance to a high percentage of the races of the pathogen behaved as expected for "horizontal" or "general" resistance.

R.A. Robinson (1971) discussed the value of vertical resistance (VR) in agriculture. VR is easy to manipulate in a breeding programme. However, it is liable to break down when the pathogen produces a new and virulent pathotype. Both the time for a breakdown and its agricultural importance vary greatly, depending on the nature of the disease in question. R.A. Robinson (1971) elaborated 14 situations as follows:

1. VR is unlikely to be valuable in a perennial crop or in a crop that is difficult to breed, because of the fact that a breakdown of resistance will be permanent. This will require replacement of the host variety, which is easy to breed with annual species but difficult with perennial species.

2. VR is likely to be more valuable against a simple interest disease(s) in which a longer time is required for a new pathotype to become widespread, for instance the soil/seed-borne pathogens; but if it is a compound interest disease, the population of a new vertical pathotype can explode across an entire continent within a season or two; for example, potato blight and cereal rusts.

3. VR is unlikely to be valuable against the pathogens which have high vertical mutability.

4. VR is unlikely to be valuable when the host population is genetically uniform and is grown in large acreages of a single cultivar. This is because, if a new vertical pathotype is produced, its increase is determined by selection pressure, which depends on both the overall size and the uniformity of the vertical pathodeme.

5. VR is more likely to be valuable if stabilizing pressure can be exploited. In the absence of a vertical pathodeme, stabilizing pressure leads to a decrease in the population of the matching vertical pathotype. Within a given pathogen species the rate of this decrease is higher with some vertical pathotypes than with others.

6. Against facultative parasites one strong gene is adequate for the exploitation of stabilizing pressure, while against obligate parasite at least two strong genes are necessary. If a pathogen is an obligate parasite, stabilizing pressure on a vertical pathotype can only occur during parasitic growth, i.e. when it is growing in a host that lacks the VR in question. This stabilizing pressure can be exploited by the cultivation of at least two vertical pathodemes; it follows that at least two strong genes must be known. If a pathogen is a facultative parasite, stabilizing pressure on a vertical pathotype can also occur during saprophytic growth (i.e. as a residual population in soil) and this means that stabilizing pressure can be exploited when only one strong gene is known (Van der Plank 1968).

7. Crop and plant patterns of VR in space are valuable chiefly against compound interest diseases, i.e. pathogens which spread rapidly. This may include crop pattern (mosaics) or plant patterns (multilines). In either case the vertical pathodeme must possess a strong gene for VR to ensure that stabilizing selection operates to its maximum extent, particularly on the "super race" able to attack all the vertical pathodemes of the pattern.

8. Crop patterns of VR in time are valuable chiefly against simple interest diseases. Replacement of one vertical pathodeme with another involves rotation of resistant genes. If different crop species are involved, one strong gene is required, but if one crop species is involved, several, strong genes will be necessary.

9. VR is likely to be less valuable against a disease transmitted by the propagating material of the host. If propagating material of the host possesses VR but is nevertheless infected, it follows that it is carrying the matching virulent pathotype. This means, in effect, that the VR has broken down before the crop is even planted.

10. VR will break down more quickly if the protection it confers is incomplete. If protection is complete against non-matching pathotypes, for example potato blight, the host population will carry no diseases whatever, and the matching vertical pathotype can arise within the same host population that will be carrying the disease at a lower level, however, than in the condition where incomplete protection occurs, for instance most cereal rusts.

11. VR is likely to be more valuable if there is closed season (i.e. winter or a long tropical dry season). If there is no closed season, there is likely to be a continuous availability of host tissue that is supporting a pathogen population. More important, if that host tissue is vertical resistant, it will be carrying a matching vertical pathotype.

12. VR is likely to be more valuable when legislative control is possible. The legislation either compels or forbids the planting of a particular vertical pathodeme or provides for seed health certification etc., and thereby increases the value of VR.

13. VR is likely to be more valuable when it is reinforced with useful levels of HR. For epidemiological reasons, VR is more effective when it is supported by HR.

14. The breakdown of a complex VR may result in less disease than the breakdown of a simple VR. There appears to be an inverse relationship between vertical pathogenicity and horizontal pathogenicity. A complex vertical pathotype is likely to have a reduced horizontal pathogenicity. A reduced horizontal pathogenicity has an effect identical to an increased HR: there is less disease.

e) **Cytoplasmic Resistance.** The cytoplasm and its organelles play an important role in heredity. Nagaich et al. (1968) reported that the inheritance of symptoms of potato mosaic virus X infection in the interspecific hybrids of *Capsicum* species is controlled by the cytoplasm of the female parent. The two species *C. annuum* and *C. pendulum* differ in showing, respectively, a systematic mosaic and necrotic local reaction after leaf inoculation. The phenotype of the F_1 hybrid depends on which parent is used as female.

The two kinds of cytoplasmic male sterility were released initially in maize. One had been discovered by Mangelsdorf and Rogers at the Texas Agricultural Experiment Station in plants of the varieties of Mexican June and Golden June. The other was discovered in corn material collected by M.J. Jenkins of the USDA and sent to Jones. The two kinds of male sterility came from different restorer genes and were designated T (Texas = Tms) and S (Sterile = Sms).

The most clearly established case of a cytoplasmically determined reaction to a fungal pathogen is the sensitivity of Tms male-sterile corn to race T of *Drechselera maydis* (Villareal and Lantican 1965). Hybrids carrying this type of cytoplasm are susceptible irrespective of whether fertility has been restored, but hybrids with other cytoplasm were not badly diseased. These other cytoplasms, designated by letters (C, S, and so on), are best distinguished from each other by their response to specific fertility-restoring genes.

Seedlings of inbred lines of corn with T or with P cytoplasm for male sterility were susceptible to *D. maydis*. Seedlings with regular (non-sterile) cytoplasm or with S or

C cytoplasm for male sterility were resistant. Differences in disease reaction were expressed in the form of lesion type, sporulation in the lesion when incubated at high humidity, inhibition of primary root elongation when incubated in a pathotoxin from the fungus, and in water soaking of leaf sections floated on the pathotoxin (A.L. Hooker et al. 1970).

Maize inbreds were developed with a back-cross programme which had either the CI 21 (A) or the GA 199 genotype with a different cytoplasm.

Seeds of maize inbred CI 21 (Athens) were prepared with the genotype of CI 21 (A) and the cytoplasms of inbreds GA 199, GT 112 and CI 21 (A) separately in back-cross programmes. Similarly, two cytoplasms of inbreds GA 199 and CI 21 were prepared with the genome of the GA 199 inbred. Thus each lot of seed had either the CI 21 (A) or the GA 199 genotype with a different cytoplasm (A.P. Rao and Fleming 1980).

The cytoplasms were classified for reaction to *Drechslera turcica (H. turcicum)* the northern blight pathogen, based on a 1 to 5 visual rating scale. Results with CI 21 (A) genome indicated a highly significant difference between GA 199 and GT 112 cytoplasms. The GA 199 cytoplasm offered more tolerance than the GT 112 cytoplasm.

A highly significant difference was also obtained between cytoplasms CI 21 and GA 199 with GA 199 nucleus. The CI 21 cytoplasm was more resistant than the GA 199 cytoplasm.

The differences demonstrate cytoplasmic influences. Therefore, it might be beneficial to screen inbred cytoplasms for resistance to northern leaf blight of maize so that cytoplasmic diversification can be practised in a breeding programme (A.P. Rao and Fleming 1980). The susceptibility in maize to yellow leaf blight caused by *Phyllosticta zeae* is governed by a cytoplasmic factor(s) conferring male sterility, which insures uniform susceptibility to yellow leaf blight of maize hybrids, even though many inbred lines with a high degree of resistance are available (Scheifele et al. 1969).

f) Resistance Based on Defense Mechanisms. This is treated in the following section.

2.6 Structural Defense Mechanisms

Some of the structural characteristics of the plant act as a physical barrier and make it difficult or impossible for the pathogen to enter the plant or to spread through it. These are of two types preformed and induced.

2.6.1 Preformed Defense Structures

These are present at the surface or in the tissues of the plant regardless of any contact the plant might or might not have had with the pathogen.

a) Role of Wax and Cuticle. Waxes are thought to play a defensive role on leaf and fruit surfaces by forming a hydrophobic surface that acts as water-repellent and prevents the retention of water drop/film of water on which pathogens might be deposited and germinate/multiply. Malformed appressoria of *Erysiphe graminis* were observed on wheat and barley plants that possessed eceriferum (eer) mutations which affect the components and physical structure of wax layers (Yang and Ellingboe 1972). A thick mat of hairs on the plant surface may also exert a water-repelling effect and may reduce infection.

Cuticle thickness has often been linked to resistance to infection in diseases in which the pathogen enters its host only through direct penetration. This may be more important in the case of some of the fungal pathogens which depend mainly on mechanical pressure for penetration into its host. Stockwell and Hanchy (1984) studied the role of cuticle in resistance of beans to *Rhizoctonia solani*. They found that increased calcification of cell walls and increased thickness were important in 3-weeks-old red kidney bean plants. A waxy cuticle may also limit exudation of the nutrients etc. required by the pathogens in the initial stages of infection and, therefore, may indirectly contribute to the defense of the plant.

In some infections the cuticle itself is by-passed via the stomata and other natural openings; where penetration occurs and excluding the possibility of pathways through it, the cuticle is often feebly developed (Cutin < 0.1 mg cm^{-2} of surface) and is of such a composition that it cannot be considered to provide a serious barrier to invasion (J.T. Martin 1964).

b) The Epidermal Layer. The epidermis is the outer and primary layer of stems, roots, leaves etc. and includes waxes and cuticle and appendages like trichomes on the aerial parts and root hairs. The thick and tough epidermal cell walls make direct penetration by the fungal pathogens difficult or impossible. Even with the same thickness, the toughness of the epidermal layer may vary because of lignification, presence of silicic acid etc. such as in rice plants, the outer walls of most epidermal cells are lignified in the resistant varieties and are seldom invaded by the rice blast fungus *(P. oryzae)*. Suberization of the epidermis provides protection against plant pathogens. The primary resistance mechanism in potato tubers to *Fusarium* dry rot *(F. roseum)* appears to be the formation of a suberin layer followed by wound periderm differentiation (O'Brien and Leach 1983).

The trichomes present on the epidermal layer can protect the plants from pathogens by repelling their vectors.

c) Structure of Natural Openings. The resistance of plants to infection pathogens which enter through the stomata depends on the behaviour of the guard cells in opening and closing the aperture etc. Some wheat varieties, in which the stomata open late in the day, are resistant because the germ tubes of spores germinating in the night dew desiccate owing to the evaporation of the dew before stomata begin to open. The mandarin variety Szinkum is resistant to the citrus canker bacterium, *Pseudomonas syringae* pv. *citri*, because the stomata of this variety have a very narrow entrance.

Penetration through lenticels is rather common in several fungal and bacterial diseases. The size and the internal structure of lenticels are important as defensive factors against disease; for instance, the small lenticels of apple varieties protect them from infection by the apple spot bacterium *(Pseudomonas papulosum)*. Hydathodes may also be important in providing protection against some of the bacterial pathogens.

d) Root Cap and Mucilage. The root cap and mucilage provide a protective covering over the outer root walls. Through lubrication, mucilage prevents excessive wounding of root surface by abrasion against soil particles, and therefore reduces the potential for invasion by soil-borne pathogens.

e) Seed Coat. The seed coat provides an effective barrier against penetration by many pathogens. It protects the embryo from dessication and from physical and biological damage. In general, mature and well-developed seeds are more resistant to diseases than poorly developed ones. H.J. Hill and West (1982) studied fungal penetration of soybean seed through pores. They postulated that mycelium of seed-borne fungi enter through seed coat defects and the hilum region in soybeans *(Glycine max)*. Using scanning electron microscopy, naturally occurring pores on the seed coat were observed as providing a means of entry into the seed. These pores were found to penetrate deeply into the palisade layer providing passage into the houseglass layer. Fungal hyphae were observed to extend into these pores, which thus provide a means of fungal entry without the presence of visible seed coat defects.

f) Internal Physical Barriers. The thickness and toughness of the cell walls of the tissues being invaded vary and may sometimes make the advance of the pathogen difficult; for example, the presence of extended areas of sclerenchyma cells may stop the spread of some pathogens like the stem rust of cereals. The xylem, bundle sheath, and sclerenchyma cells of the leaf veins effectively block the spread of some fungal, bacterial, and nematode pathogens which cause the various angular leaf spots because of their spread into areas only between but not across, veins.

2.6.2 Defense Structures Formed in Response to Infection by the Pathogen

Some of the structural characteristics are absent in the healthy plant but begin to form as soon as it is attacked by the pathogen and in response to the infection by the pathogen. These are called induced defense structures. These may be either histological, such as formation of cork layers, abscission layers, tyloses, deposition of gum etc., or cellular, such as involving morphological changes in the cell wall, etc.

a) Formation of Cork Layers. Infection of plants by some plant pathogens induces the formation of several layers of cork cells beyond the point of infection, apparently as a result of stimulation of the host cells by substances secreted by the pathogen. The cork layers not only inhibit further invasion by the pathogen, but also block the spread of any toxic substances that the pathogen may secrete. The effectiveness of cork layers depends on the speed with which it is produced following infection, on the thickness and degree of impregnation of the cork cell walls with suberin or lignin, and on the properties of the particular pathogen.

A lignin-like substance has been found in the root tissues of Japanese radish (*Raphanus sativus* var. *hortensis*) affected by *Alternaria japonica*. The methoxyl content of the isolated lignin preparation from root tissues affected by fungus was different from the native lignin of spruce. This substance was not found in the healthy parenchyma.

Living tissues affected by the fungus alone were capable of forming this substance (Asada and Matsmumoto 1967).

Beardmore et al. (1983) observed the lignification in the necrotic cells of hypersensitive response in wheat to stem rust (*P. graminis* f.sp. *tritici*).

b) **Formation of Abscission Layers.** Abscission layers are formed on young active leaves of stone fruit trees following infection by any pathogen. This layer consists of a gap between two circular layers of cells of leaf surrounding the locus of infection. Upon infection, the middle lamella is dissolved throughout the thickness of the leaf, completely cutting off the central area from the rest of the leaf. Gradually this area shrivels, dies and sloughs off, carrying the pathogen with it. Thus the plant, by discarding the infected areas along with a few uninfected cells, protects the rest of the leaf tissue from invasion by the pathogen. In peach leaves, the abscission layer is formed in response to infection by the bacterium, *Xanthomonas pruni*.

c) **Formation of Tyloses.** Tyloses are outgrowths of the protoplast of adjacent living parenchymatous cells which protrude into xylem vessels through half-bordered pits. Tyloses have cellulosic walls and may, by their size and numbers, clog the vessel completely. The time and rapidity of tylose formation determine whether its role will be defensive or whether it will be one of the factors for causing disease. In sweet potato wilt (*Fusarium oxysporum* f. *batatas*), tyloses in some varieties are formed abundantly and quickly before pathogen penetration, thus bringing about resistance by preventing the spread of the pathogen.

d) **Deposition of Gums.** Various types of gum are produced by many plant around lesions following infection by pathogens or injury through mechanical means or insects etc. The defensive role of gums is in that they form quickly in the intercellular spaces and within the cells surrounding the locus of infection, thus forming an impenetrable barrier around the pathogen which becomes isolated, and sooner or later dies. Such a mechanism for resistance operates in rice blast *(P. oryzae)* and leaf spot *(Helminthosporium oryzae = Drechslera oryzae)*.

e) **Silicon Content.** A quantitative study has been made of the relationship between the silicon content of leaves of Caloro rice and their susceptibility to infection by *P. oryzae* Cav. (Volk et al. 1958). Silicon content and susceptibility are related inversely. Both silicon content and the degree of susceptibility of the leaf at any moment are related to the amount of silicate available to the roots. It is hypothesized that the silicon combines with one or more components of the cell wall to form a complex relatively resistant to attack by the extracellular enzymes of *Pyricularia*, thus diminishing hyphal penetration into the leaf.

f) **Swelling of Cell Wall.** Swelling of epidermal and sub-epidermal cells during infection by the pathogen in direct penetration of host may inhibit invasion and establishment of infection by the pathogen. The resistant varieties of cucumber to scab fungus, *Cladosporium cucumerinum*, produces an inhibitor which is capable of inactivating the pectinolytic enzymes secreted by the pathogen. The pectinase separates the cells, and the activity of this enzyme continues/unchecked in susceptible tissues coupled with a destruction of inhibitor; however, in resistant tissues the growth of pathogen is inhibited (Mahadevan et al. 1965). The depositions of callose by plant cells have been shown to prevent successful penetration in papaya fruit by *Colletotrichum gloeosporioides* (Stanghellini and Aragaki 1966).

g) **Sheathing of Hyphae.** Hyphae of fungi penetrating a cell wall are often enveloped in a sheath formed by the extension of the cell walls and prevent the pathogen's spread. The sheath consists of cellulose, callose or other materials, or may be the deposition product of the cytoplasm rather than the cell wall.

h) **Cytoplasmic Defense Reaction.** The earliest cytoplasmic response to contact between a microorganism and a host cell is cyclosis within the protoplast, with a resultant reorganization of cytoplasmic organelles. Infection of tuber tissues and tissue culture cells of *Solanum tuberosum* by *Phytophthora erythroseptica*, *P. infestans* and *Fusarium caeruleum* caused swelling and disruption of host cytoplasmic particles containing acid phosphatases, esterases and proteases. Heavy diffuse cytoplasmic staining for acid phosphatase was a consistent feature of infection by all three fungi, but staining reactions for esterase and proteases showed much less diffuse staining and a lesser degree of particle swelling. An excess recovery of ribonuclease from tissues infected with *P. infestans* and *F. caeruleum* was found (Pitt and Coombes 1969). An electrophoretic comparison of near-isogenic lines of wilt (*Fusarium oxysporum* f.sp. *pisi*) resistant or susceptible *Pisum sativum* was made by Hunt and Barnes (1982). The resistant lines differed from susceptible lines in carrying an esterase component which has been derived from the resistant cultivar Delwiche Commando.

i) **Hypersensitive Reaction.** This is one of the most important defense mechanisms in plants. It occurs only in incompatible host-pathogen combinations where the pathogen may penetrate the cell wall, but as soon as it establishes contact with the protoplast of the cell, the nucleus moves towards the intruding pathogen and soon disintegrates, leading to loss of turgor and formation of brown resin like granules in the cytoplasm. Such changes do not occur in susceptible varieties or may occur at a much slower rate. In most of the incompatible cases the hypha does not grow out of such cells and further invasion is stopped. The pathogens within the area of operation of hypersensitive reaction are isolated by necrotic tissue and quickly die. Necrosis is normally characterized by formation of brown to black pigments (melanin) throughout the cell walls and the collapsed protoplast walls of adjoining live cells also may be melanized.

2.7 Biochemical Defense Mechanisms

In resistant varieties in which biochemical defense mechanisms are operating, the growth rate of disease lesions soon slows down and finally, in the absence of structural defenses, their growth is completely checked. The biochemical defense mechanisms could be pre-existing or induced in response to attacking pathogens.

2.7.1 Pre-Existing Biochemical Defense

In such mechanisms certain chemical compounds are already present in the resistant host varieties and interfere with the growth and multiplication of the pathogen. These are of the following kinds.

a) **Inhibitors Released by the Plant in its Environment.** Some of the plants release substances through the surface of aboveground as well as underground parts which are fungitoxic and hence inhibit germination of fungal spores. Fungitoxic exudates on the leaves of tomato and sugarbeet, for instance, seem to inhibit germination of conidia of *Botrytis* and *Cercospora*, respectively, in dew or rain droplets on these leaves. Resistant flax varieties exude a glucoside which upon breakdown produces hydrocyanide (HCN), which is an extremely potent poison to living organisms, including *F. oxysporum* f. *lini*. The growth of this fungus was inhibited in vitro at 135 ppm HCN (Trione 1960).

b) **Inhibitors Present in Plant Cells Before Infection.** In the case of the potato scab disease caused by *Streptomyces scabies*, tubers of resistant varieties contain higher concentrations of chlorogenic acid in the lenticels than do the tubers of susceptible varieties. Chlorogenic acid is a phenolic compound toxic to the pathogen.

The higher content of chlorogenic acid in the roots of resistant potato varieties is also considered as the main mechanism of defense against *Verticillium* wilt.

c) **Defense Through Deficiency in Nutrients Essential for the Pathogen.** The host specialization of the pathogen may be due to the specific need of these pathogens for a substance that is present in adequate quantities only in the host(s) they can infect. Therefore the species or varieties that do not produce this substance would be resistant to the pathogen that requires it.

d) **Defense Through Common Antigens.** When a variety does not have an antigen that is present in a particular rust race, the variety is resistant to that race, suggesting that susceptibility or resistance are due to the presence or absence of the specific rust antigens. Antigens obtained from crude host tissue extracts of alfalfa *(Medicago sativa)* did not react with bacterial *(Corynebacterium insidiosum)* antisera, which demonstrated the lack of common antigens between the host and pathogen (Carroll et al. 1972). Such a mechanism also operates in flax rust caused by *M. lini* and angular leaf spot of cotton caused by *X. campestris* pv. *malvacearum*.

2.7.2 Induced Biochemical Defense

Such substances are produced in response to injury as well as to infection. However, these are produced in higher quantities following infection rather than injury, probably because of greater physiological stress due to the continuous irritation of the infected tissue by the pathogen. These compounds are mainly phenolic. The phenolic compounds are of two types.

a) **Common Phenolics**. Some of the phenolic compounds are found in healthy as well as in diseased plants, but their synthesis or accumulation seems to be accelerated following infection. Such compounds are called common phenolics. These are toxic to the pathogens and their production and accumulation proceed at a faster rate after infection in a resistant variety than in a susceptible variety.

The spread of phenol accumulation in the infected tissues was correlated with infection type in wheat against stem rust, resistant reaction being associated with a consistently faster accumulation of phenolics (A. Kiraly and Farkas 1962).

A positive and striking correlation was found between peroxidase and polyphenoloxidase activity and the resistance of potato tissues to *Phytophthora infestans* (Fehrmann and Dimond 1967). No such correlation could be obtained for the chlorogenic acid content of the tissues. Seevers and Daly (1970) measured peroxidase activity induced by *P. graminis* f.sp. *tritici* in wheat and reported that it does not appear to be correlated with resistance or susceptibility as measured by infection type. Although indolacetic acid (IAA) or its metabolism may not be concerned directly with host-parasite compatibility, the degradation of exogenous IAA appears to be under the influence of metabolic systems that are connected, directly or indirectly, with resistance or susceptibility (Antonelli and Daly 1966).

The peroxidase and orthodiphenol oxidase increased in *Beta vulgaris* in response to infection against *Cercospora betae*, the leaf spot pathogen, and the increase was consistently higher in resistant than in susceptible varieties. Towards the advanced stages of the disease, this pattern was reversed, and the increase in the amounts of enzymes was higher in susceptible varieties (Rautela and Payne 1970).

The plants of resistant variety of rice, TKM6 inoculated with *X. campestris* pv. *oryzae* had larger quantities of total ortho-dihydroxyphenols than both Co13 and IR 8 the susceptible varieties (Purushothaman 1975). Purkayastha and Raha (1983) isolated an antifungal substance from jute (*Corchorus* spp.) leaf extract and it was identified as chlorogenic acid. This phenolic acid component increased in jute leaves after *Myrothecium roridum* infection, more so in leaves of infected resistant cultivars. The rate at which chlorogenic acid accumulates in leaves after infection is important in determining resistance in jute.

b) **Phytoalexins**. These are also phenolic compounds; they are, however, not present in healthy plants, but are produced upon stimulation of plants by micro-organisms or by chemical or mechanical injury. These inhibit the growth of plant pathogens. The concept that plants produce protective chemicals after infection was formalized by Müller and Börger (1941). However, their chemical entity as phytoalexins was shown by Müller (1958) while working with the hypersensitive response of bean tissue to the soft-fruit pathogen, *Monilinia fructicola*. The field of phytoalexins has been studied intensively in the past years and a number of recent reviews are available (Grisebach and Ebel 1978, Gross 1977, Kuc 1976, Van Etten and Pueppke 1976, Kuc et al. 1976 and Deverall 1976). The phytoalexins in some tribes of Leguminosae are presented in Table 2.2 and of Solanaceae and tribes other than Leguminosae in Table 2.3. The action of some of these is described in detail in the ensuing paragraphs.

Pisatin, an antifungal compound, was isolated from hypersensitivity studies using pods of *Pisum sativum* (Table 2.2) in response to infections with *Monilinia fructicola* (Cruickshank and Perrin 1960).

Table 2.2. Phytoalexins in some tribes of the leguminosae. (After Deverall 1977)

Tribe	Species	Phytoalexins	Reference
Dioeleae	*Canavalia ensiformis*	Medicarpin	Keen (1972)
Phaseoleae	*Phaseolus vulgaris*	Pisatin	Sanz (1981)
		Phaseolin	Perrin (1964), Sanz (1981),
		Phaseolidin	Perrin et al. (1972),
		Phaseolinisoflavan	R.S. Burden et al. (1972)
		Kievitone	Anderson-Prouty and
			Albersheim (1975),
			A.J. Anderson (1978),
			Fraile et al. (1982)
	P. lunatus	Phaseolin	Cruickshank and Perrin (1971)
	P. radiatus		
	P. leucanthus		
	Vigna unguiculata	Phaseolidin	J.A. Bailey (1973)
		Kievitone	
		Phaseolin	
Glycineae	*Glycine max*	Glycoprotein	Frank and Paxton (1971),
		A soybean	R.S. Burden and J.A. Bailey
		Pterocarpan	(1975), Albersheim and Valent
		Glyceollins	(1978)
Vicieae	*Cicer arietinum*	Medicarpin	Keen (1975)
		Maackiain	
	Vicia faba	Wyerone acid	Letcher et al. (1970),
		Wyerone	Fawcett et al. (1971),
			Hargreavens et al. (1977),
		Medicarpin	Ibrahim et al. (1982)
	Pisum sativum	Pisatin	Perrin and Bottomley (1962),
			Hadwiger and Beckmann (1980),
		Maackiain	Stoessl (1972)
	P. arvense	Pisatin	Cruickshank and Perrin (1960,
	P. elatius		1965)
	P. abyssinicum		
	P. fulvum		
Trifolieae	*Medicago sativa*	Medicarpin	D.G. Smith et al. (1971),
			Vaziri et al. (1981),
		Sativan	Ingham and Miller (1973)
	Trifolium pratense	Maackiain	Higgins and D.G. Smith (1972)
		Medicarpin	
	T. repens	Medicarpin	Cruickshank et al. (1974)
Loteae	*Lotus corniculatus*	Sativin	Bonde et al. (1973)
		Vestifol	

The free amino acids and amides in healthy and infected leaf tissue of halo blight-resistant and susceptible plants showed distinct differences (Patel and Walker 1963). In susceptible plants, there was an increase in ornithine, histidine, methionine, asparaginge glutamine, β-alanine and lysine in inoculated leaves. Chlorotic leaves on inoculated plants to which there was limited movement of the pathogen from the site of infection showed even greater increases. In resistant plants, there was little difference between the healthy and infected plants.

Table 2.3. Phytoalexins in plant families Solanaceae and other than Leguminosae. (After Deverall 1977)

Family	Species	Phytoalexins	Reference
Chenopodiaceae	*Beta vulgaris*	2,5-Dimethylene-dioxyflavanone 2'-Hydroxy-5-methoxy-6,7-methylenedioxyiso-flavone	Geigert et al. (1973)
Malvaceae	*Gossypium barbadense*	Vergosin Hemigossypol	Zaki et al. (1972)
Umbelliferae	*Daucus carota*	3-methy-6-methoxy-8-hydroxy-3,4-dihydro-isocoumarin	Codon and Kuc (1962)
	Pastinaca sativa	*Xanthotoxin*	C. Johnson et al. (1973)
Convolvulaceae	*Ipomoea batatas*	Ipomeamarone	Kubota and Matsuura (1953)
Compositae	*Carthamus tinctorius*	Safynol Dehydrosafynol	Allen and Thomas (1971b)
Orchidaceae	*Orchis militaris*	Orchinol	Hardegger et al. (1963)
	Loroglossum hircinum	Hircinol	Gäumann (1964)
Solanaceae	*Solanum tuberosum*	Rishitin (terpenoid compounds)	Tomiyama et al. (1968), Katsui et al. (1968), Elgersma (1980), Glazener and Wouters (1981), Sanz (1981)
		Lubinin Phytuberin	Metlitskii et al. (1971) Varns et al. (1971)
	Lycopersicon esculentum	Falcarindiol (cis-heptadeca-1,9-diene-4,6-diyne-3,8-diol) cis-tetra-deea-6-ene 1,3-diyne-5, 8-diol	De Wit and Kodde (1981)
Scitaminae	*Costus speciosus*	Glyceollin	Kumar et al. (1984)

Rishitin may play some role in the defense reaction of potato tubers to infection by *P. infestans* fungus (Table 2.3). The fact that rishitin acts as a growth retardant suggests that it may affect not only the parasite but also the host plant tissue (Tomiyama et al. 1968).

Resistance in soybean against *Phytophthora megasperma* var. *sojae* was at least of two types; resistance in young plant tissue (0-2 weeks old) in which the production of phytoalexins played an important role and resistance in older plant tissue (> 2 weeks old) recognized by woody stem tissue and greatly reduced phytoalexin (Paxton and Chamberlain 1969). An inducer capable of inducing phytoalexin production in soybean plants was isolated from *P. megasperma* var. *sojae* (Table 2.2). The chemical tests indicated that it was glycoprotein (Frank and Paxton 1971).

The limited growth of fungus in the immune host-pathogen combination of *Z. mays* and *P. graminis* apparently resulted from the action of a phytoalexin (K.T. Leath and Rowell 1970).

Monogenic resistance in corn to pathogenic isolates of *D. turcica (= H. turcicum)* is probably chemical and two phytoalexins A_1 and A_2 were found in monogenic resistant corn (Lim et al. 1970). The production of phytoalexin is conditioned by the gene Ht (Lim et al. 1968) and is induced to form in resistant corn when host pathogen interact.

Two antifungal compounds namely safynol and dehydrosafynol (Table 2.3) were found in high concentrations in tissues exterior to the vascular ring in response to *Phytophthora drechsleri* in safflower, *Carthamus tinctorius* (Allen and Thomas 1971b).

Bean *(Phaseolus vulgaris)* hypocotyles, pea *(Pisum sativum)* pods and tomato *(Lyco-persicon esculentum)* fruits were tested for phaseolin, pisatin and rishitin production when challenged with the phytopathogenic bacteria *Erwinia carotovora, Pseudomonas syringae* pv. *phaseolicola, P. syringae* pv. *pisi* and *P. solanacearum*, and their isolated extra-cellular polysaccharides. All bacteria induced phytoalexin accumulation, whereas only phaseolin and pisatin, but not rishitin, were elicited by extracellular polysac-charides. The inhibitory effect of these three phytoalexins on bacterial growth was studied in liquid medium, whereas phaseolin and pisatin strongly inhibited growth, only a slight inhibitory effect resulting from the presence of rishitin in the medium (Sanz 1981).

3 Concepts in Insect-Pest Resistance

Insects are usually specialized in their ability to attack their hosts or their parts. No insect is capable of attacking every species of hosts. Even the so-called polyphagous insect cannot feed on just any plant species. Even the oligophagous insects are restricted in their ability to utilize the varieties of one family of the host plant. Early literature contains several significant examples of differences in response to insect attack. The first mention of the possibility of resistance to the attack of hessian fly *(Mayetiola destructor)* was the Underhill variety of wheat in the year 1785. However, the first observations were published by Lindley (1831), Wickson (1886), Woodsworth (1891), and Kellner (1892). Snelling (1941b) reviewed the literature on this subject and observed that out of 567 references only 37 papers prior to 1920 dealt with plant resistance to insects. The publications in this field increased considerably after this period. According to him, plant resistance includes those characters which enable a plant to avoid, tolerate or recover from the attacks of insects under conditions that would cause greater injury to other plants of the same species.

Painter (1951) wrote the first book on insect resistance in crop plants, and made agricultural scientists cognizant of the fact that the use of resistant crop plant was an ideal way to protect crops against insect pests. According to him, plants that are inherently less damaged/infested than others under comparable environmental conditions in the field are resistant (Painter 1958). Beck (1965) defined plant resistance as "being the collective heritable characteristics by which a plant species, race or individual may reduce the probability of successful utilization of that plant or a host by an insect species, race, biotype or individual". Maxwell et al. (1972) defined resistance as "those heritable characteristics possessed by the plant which influence the ultimate degree of damage done by insects".

3.1 Degree of Resistance

Resistance is relative and is measured by using susceptible cultivars of the same plant species as checks. The degree of resistance among specific host plants may oscillate between two extremes; immunity and high susceptibility. An immune plant is a non-host. Any degree of host reaction less than immunity is resistance; more than immune is impossible. Painter (1951) used the following scale to classify degrees of decreasing resistance:

a) **Immunity.** An immune cultiver is one that a specific insect will never consume or injure under any known condition. Immunity to *Amphorophora agathonica*, the aphid vector of red raspberry mosaic virus, was reported (Daubeny 1972).

b) **High Resistance.** A variety with high resistance is one which possesses qualities resulting in small damage by a specific insect under a given set of conditions.

c) **Low Resistance.** A low level of resistance indicates the possession of qualities which cause a variety to show less damage or infestation by an insect than the average for the crop under consideration.

d) **Susceptibility.** A susceptible variety is that which shows average or more than average infestation or damage by an insect.

e) **High Susceptibility.** A variety shows high susceptibility when much more than average damage is done by the insect under consideration.

3.2 Functional Resistance

Certain phenomenon related to resistance but not necessarily based on heritable traits were defined and classified by Painter (1951).

The term pseudo-resistance is not the result of genetic characters inherent in the plant, but results from some temporary shifts in the environmental conditions favourable to the otherwise susceptible host plant. Cultivars or species showing pseudo-resistance are important in economic entomology, but should be distinguished from cultivars that show resistance throughout a wider range of environments, i.e. with inherent mechanisms. The functional resistance can be classified into three types.

a) **Host Evasion.** For an insect pest, not only finding the right host, but also contacting it at an appropriate stage of development are of equal importance. The phenologies of the host and insect must synchronize. Host evasion takes place when the plant growth pattern is modified so as to bring in asynchronies of insect-host phenologies (Panda 1979). Early maturity has been used to be a good example in economic entomology. Rice crop planted in late August and thereafter was severely affected by gall midge *(Pachydiplosis oryzae)* in Orissa, India, because the peak period of gall midge infestation was from the second week of October to the second week of November. The transplantation time has thus a direct influence on the prevalence and incidence of this pest (Prakasa Rao et al. 1971). Therefore, early flowering could provide the best way to evade the attack by the pest. In *Sorghum*, the shoot fly *(Atherigona varia soccata)* was almost absent at the beginning of the monsoon season, and early plantings were completely free from shoot-fly incidence in India (Vidyabhushanam 1972).

b) **Induced Resistance.** This term may be used for temporarily increased resistance resulting from some condition of plant or environment such as amount of water and soil fertility etc. Aphids are particularly sensitive to the levels of nitrogen in the plant, but respond negatively to the levels of K, even in the presence of high nitrogen (Van Emden 1966). The relationship between total nitrogen, sugars and insect damage, with few exceptions, was observed in maize and sorghum against *Chilo zonellus* (Kalode and Plant 1967). Rajaratnam and Hock (1975) reported highest mite concentration in oil palm seedlings which had received little or no boron in the growing media. When deficient seedlings were supplied with boron, the new leaves were less severely affected by mites than the older leaves.

A decrease in ascorbic acid obtained by application of lycorine (an inhibitor of ascorbic acid synthesis) induces a reduction of plant resistance to the nematode *(Meloidogyne incognita)*, but an artificial increase in ascorbic acid concentrations transformed susceptible plants into resistant ones. The amount of ascorbic acid in susceptible plants was unaltered by nematode attacks, but in resistant plants ascorbic acid synthesis was always stimulated (Arrigoni et al. 1979). The natural resistance of soybean flour to *Trogoderma granarium* is mainly due to two factors, namely heat-labile inhibitor and nutritional deficiency (Chaudhary and Bhattacharya 1982).

Such induced resistance may be useful, but should not be confused with inherent differences in resistance between cultivars or plants.

c) **Escape.** This refers to the lack of infestation or injury to the host plant because of such transitory circumstances as incomplete infestation. Thus, finding an uninfested plant in a susceptible population does not necessarily mean that it is resistant. Even under very heavy infestation susceptible plants will occasionally escape, so that only studies of their progenies will establish their true relationship (Horber 1979).

3.3 Types of Genetic Resistance

Host-plant resistance may be caused as a result of a series of interactions between insects and plants which influence the selection of plants as hosts and the effects of plants on insect survival and multiplication. An empirical approach was proposed by Painter (1936, 1941, 1951), that proved a workable compromise between mere categorization of phenomenon and the basic study of causative factors or processes. Painter (1958) proposed mechanisms of resistance which were grouped into three main categories: non-preference, antibiosis and tolerance (Fig. 3.1). Hanover (1975), however, postulated four basic plant resistance mechanisms; (1) the morphology and anatomy of host; (2) chemical repellents produced by the host; (3) chemical attractants produced by the host; and (4) the nutritional status of host. Here, the non-preference, antibiosis and tolerance classification of mechanism of resistance by Painter is discussed along with the good points of Hanover's four-point classification.

a) **Preference and Non-Preference.** Non-preference is the insects' response to plants that lack the characteristic to serve as hosts, resulting from negative response or total avoidance during search for food, oviposition sites or shelter or for the combinations of the three. Kogan and Ortman (1978) proposed substituting "antixenosis" for "non-

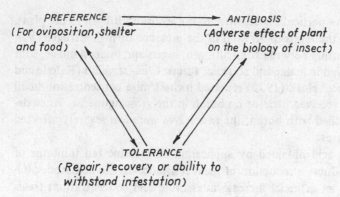

Fig. 3.1. Three interrelated characteristics, one or more of which is frequently present in resistant varieties (Painter 1951)

preference". It is a term parallel to antibiosis, and conveys the idea that the plant is avoided as a "bad host".

Panda (1979) reported that at least two types of non-preference mechanism have been observed: one which can be manifested only in the presence of the preferred host, the second even in the absence of the preferred host; such a mechanism is strong enough to force the insect to starve, particularly in early instar stages. Therefore, the extreme non-preference may be mistaken for antibiosis and vice-versa.

Brazzel and Martin (1957) studied the number of eggs laid by pink boll worm, *(Pectinophora gossypiella)* on *Gossypium thurberi* in comparison with the number laid on the-susceptible cotton variety Deltapine 15, of *G. hirsutum*. When the two kinds of plant were confined together in the same cage, 79% of the *G. thurberi* bolls were without eggs, while 50% were without eggs when *G. thurberi* was alone in the cage.

The following factors govern the preference and non-preference:

— Insect behaviour as a basis for preference.
— Preference differences as a response to colours or intensity of light.
— Factors conditioning preference.
— Preference differences as responses to mechanical stimuli from the physical structure and surface of plants.
— Preference differences as responses to chemical stimuli.
— Possibilities of preference as a resistant mechanism.

Establishment of insects on plants involves orientation, feeding, metabolic utilization of ingested food, growth, survival and egg production (K.N. Saxena 1969, K.N. Saxena et al. 1974) and hatching (K.N. Saxena and Pathak 1977). Of these, orientation, feeding and oviposition are behavioral responses and the others are physiological responses. The interruptions in the different phases of insect establishment may be caused by biophysical or biochemical characteristics, but it is often difficult to distinguish between the effects of physical and chemical factors on insects, as similar results may come from diverse causes.

The yellow-green varieties of pea *(Pisum sativum)* are more resistant to the pea aphid than the blue-green varieties (Searls 1935, Cody 1941). Fewer aphids were found on the yellow-green varieties. At the seedling stage (1 to 6 inches tall) colour of foliage of peas was correlated with the alighting response of winged pea aphids [*Acyrthosiphon pisum = Macrosiphum pisi*]. Migrants and colonies were found in greatest numbers on a variety with yellowish green foliage, in lesser number on varieties with green foliage, and least abundantly on a variety with deep green foliage (Cartier 1963). Red cabbage, *(Brassica oleracea)* was less preferred in both field and greenhouse by cabbage worm *Pieris rapae* for oviposition than other types tested (M.K. Dickson and Eckenrode 1975).

Morphological (physical) resistance factors interfere physically with locomotor mechanisms, and more specifically with the mechanisms of host selection, feeding, ingestion, digestion, mating, and oviposition, as opposed to those affecting the chemically mediated behavioral and metabolic processes. The physical barriers or deterrents to insects such as trichomes, surface waxes, silication or sclerotization of tissues are, however, expression of genetically regulated biochemical processes. Some of the physical resistance factors most commonly found among crop plants and their effects on insect are presented in Table 3.1. In addition, allomones affecting insect behavioral and metabolic processes may occur in plant morphological structures (trichome and bracts). Thus chemical and morphological resistance factors intertwine in a continuum of defense.

Solid-stemmed varieties of wheat have facilitated the successfull control of wheat stem sawfly *(Cephus cinctus)* (Eckroth and McNeal 1953, N.D. Holmes and Peterson 1962). Stem solidness is caused by development of pith inside the stem. When a variety is less solid, it is less resistant. Environment influences the degree of development of pith in a variety (Roberts 1954, McNeal et al. 1957). These authors pointed out that wheat varieties with the same degree of visual stem solidness may vary in their relative expression of sawfly resistance. Lebsock and Koch (1968) reported heritability values for stem solidness of 60-95%, and suggested that plant breeders should be able to select high-yielding solid-stemmed lines. The F_1 plants of wheat hybrids had stem-solidness values intermediate between those of the two parents, and only when both

Table 3.1. Physical resistance factors most commonly found among crop plants. (After Norris and Kogan 1979)

Plant factors	Effect(s) on insect
Thickening of cell walls; increased toughness of tissues	Interference with feeding and oviposition mechanisms
Proliferation of wounded tissues	Insects killed after initial injury
Solidness and other characteristics of stems	Obstruction of feeding and oviposition mechanisms, dehydration of eggs
Trichomes	Effect on feeding, digestion, oviposition, locomotion, and attachment; toxic and disruptive effects of allelochemics in glandular trichomes; provision of shelter
Accumulation of surface waxes	Effect on colonization and oviposition
Incorporation of silica	Abrasion of cuticle; feeding inhibition
Anatomical adaptations of non-specialized organs and protective structures	Various effects

parents were solid were sawfly cuttings satisfactorily controlled in F_1 plants. This suggested that for developing stem sawfly-resistant hybrid wheat, both parents must carry a solid stem (McNeal et al. 1971). A reduction in the amount of light received by the growing plants accompanied by a loss in stem solidness was the major factor responsible for the loss in sawfly resistance of bread wheats under field conditions (N.D. Holmes 1984).

Weiss (1943) reported that the responses of insects to different wavelengths may in reality result largely from response to differences in the intensity of light or to differences in the absorption of light by the primary photo-sensitive substance of the cells of the visual sense.

Plant hairs are reported to be associated with resistance to insects in at least 18 plant genera, including most major crops (Webster 1975). Poos and Smith (1931) reported that potato leafhopper *(Empoasca fabae)* heavily infested and seriously injured the glabrous soybeans, whereas the rough hairy varieties were relatively free from leafhoppers and the symptoms of leafhopper damage. Genung and Green (1962) emphasized growing soybean varieties with dense hairiness of foliage, which manifested non-preference to the female leafhoppers for oviposition.

Afzal and Abbas (1943) and Parnell et al. (1949) reported a very close and consistent relationship of hairiness of the cotton and resistance to jassids *(Empoasca* spp.). Without exception, all distinctly hairy types have been found highly resistant and all non-hairy types fully susceptible. Intermediate degrees of hairiness were associated with intermediate degrees of resistance (Parnell et al. 1949). They found such a relationship between varieties/plants of the *G. hirsutum* and *G. barbadense* species and also in the segregating progenies of the hybrids between the two species. The lack of hairiness in the early stages of growth was associated with a lack of resistance. Hairiness and resistance to jassids develop concurrently. Length of hairs was shown to be of prime importance and densities without adequate length were ineffective.

Two pairs of genes, H_1^P and H_2^A, appear to play a part in the genetic control of pubescence of leaves in cotton, *G. hirsutum*. H_1^P seems to induce hair of sufficient length and density, and is incompletely dominant to h_1^P. The H_2^A allele seems to induce hairiness but to a smaller degree. It acts additively to H_1^P, giving profusely hairy plants. H_1^P gene is present in Pak 51, L 11, and AC 134 and confers jassid resistance, while Empire Red Leaf and Acala 1517 D possess H_2^A which gave sparsely hairy plants. The variety Deltapine Smooth Leaf, possesses another gene E^A which displays an epistatic effect on H_2^A gene. The gene E^A has only a minor effect on the H_1^P. The presence of certain modifier genes seems to affect the density and length of hair, resulting in deviations even in individual groups (Muttuthamby et al. 1969).

Effects of pubescence on pink boll worm *(P. gossypiella)* damage were studied in strains of cotton *(G. hirsutum)*. Damage was lower in TM 1, a densely pubescent strain carrying the pilose allele H_2, than in most other strains tested, but levels of pubescence below that of TM 1 did not insure resistance (R.L. Wilson and F.D. Wilson 1977).

Pods of mustard *(Brassica hirta)* which have stiff hairs showed no significant flea beetle *(Phyllotreta cruciferae)* damage, while adjacent plots of rapeseed showed heavy pod damage. Removal of hairs from mustard pods caused an increase in feeding damage by the flea beetle (Lamb 1980).

The damage done by the boll worm *(H. zea)* in cotton was compared to the glanded and glandless version of 12 diverse lines (Oliver et al. 1970). The data on oviposition and damage to squares and bolls by larvae showed that the differences between the two versions were significantly different. Glabrous cottons were reported to suppress the population of cotton boll worm and tobacco bud worm *(H. virescens)* by reducing the total number of eggs deposited by adults (Lukefahr et al. 1971, Shaver and Luke-fahr 1971 and S.A. Robinson et al. 1980). However, the glabrous condition causes greater susceptibility to thrips (Lee 1971) and certain allelic combinations increase smoothness whereas others do not, and certain alleles could be linked to genes conditionally agronomically undesirable traits. Baltazer 3 variety of cotton was found to be more resistant to leafhopper, probably because of its hairy underleaf. Other non-hairy varieties were susceptible (Azawi and Campos 1974). Gland density and gossypol content in *G. hirsutum* were negatively correlated with larval seedling preference and larval weight of tobacco bud worm *H. virescens* (F.D. Wilson and Shaver 1973).

In tomatoes, a correlation between resistance of spider mites and the concentration of glandular hairs on the leaves was observed (Stoner et al. 1968). Glandular hairs on leaves and stems stuck onto naturally occurring aphids in field plants of *Solanum polyadenium* and had fewest free aphids (Gibson 1976a). Glandular hairs of this species of potato also provide a form of resistance to larvae of colorado beetle *Leptinotarsa decemlineata* (Gibson 1976b).

The correlation between leaf mid-rib structure and resistance to jassids *(Empoasca devastans)* was studied in ten varieties/interspecific and intraspecific hybrids of cotton in field (Yadava et al. 1967). High and positive correlations were found between nymphal population of jassids and thickness of cortex, and also between nymphal population and diameter of mid. veins. This suggested that the strains having thinner cortex and consequently thinner diameter of the leaf veins are resistant to jassids.

The resistance of sweet corn lines to corn ear worm *(H. zea)* has been attributed to the mechanical protection due to husk elongation and thickness and balling of the silks in the silk channel (Stoner 1970). A highly significant positive correlation between resistance to corn ear worm injury and husk tightness was observed (N.W. Widstrom et al. 1975). Differences among the lines of maize to egg-laying per plant of the European corn borer *(Ostrinia nubilalis)* were highly significant. The sweet, non-pubescent, and early lines showed few eggs per plant (Andrew and Carlson 1976).

Wheat leaf pubescence was reported to confer resistance to cereal leaf beetle *(Oulema melanopus)*, Gallun et al. (1966) and to hessian fly *(M. destructor)*, J.J. Roberts et al. (1979). Significant correlation between larval weight of cereal leaf beetle and pubescence density in common wheat *(Triticum aestivum)* revealed that the high density was associated with resistance (Ringlund and Everson 1968). The pubescence of three Purdue wheats was responsible for over 94% reduction in the number of eggs laid by cereal leaf beetle (Gallun et al. 1973). The resistance to oviposition was due to greater hair density and hair length in wheat (Webster et al. 1975). Oviposition by the beetles was less on the pubescent wheat variety Vel (CI 15890) than on the Genesee (susceptible) both in pure and mixed plots (Casagrande and Haynes 1976). Pubescence density in *Triticum turgidum* was governed by three dominant genes. The genes might be operating in additive manner to determine the length of pubescence (Leisle 1974).

Starks and Merkle (1977) noted that dense and long trichomes did not necessarily result in resistance to green bugs *(Schizaphis graminum)* in wheat.

At early growth stage, increased plant height in peas had a barrier effect to the winged migrants of pea aphid *(A. pisum)* that favoured a higher infestation by migrants.

At full growth stage, increased varietal plant height had a reduced exposure effect to ensuing aphid population because of taller stems, longer internodes and lesser dense foliage (Cartier 1963).

Successful penetration of the pods by adults of cowpea curculio *(Chaleodermis aeneus)* was negatively correlated with pod wall thickness in southern peas (Cuthbert and Davis 1972).

In alfalfa, varietal forage yields and damage score under a severe field infestation of pea-aphids *(A. pisum)*, were significantly correlated with seedling resistance in glass-house (Hackerott et al. 1963). The aphids *Myzus persicae* and *Aphis fabae* settled more readily on sugarbeet plants than on others and observed differences in settling between individual plants were mainly due to inherited characters (Russell 1966).

Non-preference was observed as the primary mechanism for resistance to the aphids *Therioaphis maculata* and *T. richmi* in alfalfa and sweet clover selections (Kishaba and Manglitz 1965); in red raspberry against the aphid *Amphorobora agathonica* (Kennedy and Schaefers 1974b); and in Brussels sprouts (J.A. Dunn and Kempton 1976). Non-preference as a mechanism of resistance in some of the crop plants is presented in Table 3.2.

Table 3.2. Non-preference as a mechanism of resistance in some of the crop plants

Crop	Insect pest	Reference
Cotton		
5 Frego lines	Cotton boll weevil *(Anthonomus grandis)*	J.N. Jenkins et al. (1969)
4 Red lines	Cotton boll weevil	J.N. Jenkins et al. (1969)
Several S.I. lines	Cotton boll weevil	J.N. Jenkins et al. (1969)
Asiatic spp.	Cotton boll weevil	Marke and Meyer (1963)
Frego bract and red	Cotton weevil	J.E. Jones et al. (1964)
Presence of hairs (H_2)	Cotton boll weevil	Stephens (1959)
Red hairy	Cotton boll weevil	Stephens and Lee (1961)
MU-9 (H_1 and H_2)	Cotton boll weevil	Wannamaker (1957)
Pilose (H_1 and H_2)	Cotton boll weevil	Wannamaker (1957)
H_1, H_2, R_1	Cotton boll weevil	Wessling (1958b)
Rice	Stem borers *(Chilo and Tryporyza* spp.)	Pathak (1977)
Wheat	Stem sawfly *(Cephus cinctus)*	Platt et al. (1941)
	Cereal leaf beetle *(Oulema melanopus)*	Ringlund and Everson (1968)
Soybean	Mexican bean beetle *(Epilachna varivestis)*	Schillinger (1976)
Potato		
(Glandular trichomes)	Green peachaphid *(Myzus persicae)*	Tingey and Laubengayer (1981)
	Potato leafhopper *(Empoasca fabae)*	
Alfalfa		
(Glandular haired species)	Spotted alfalfa aphid *(Therioaphis maculata)*	Ferguson et al. (1982)

Table 3.3. Allelochemic factors in plants and their effects on insects corresponding behavioral or physiological response. (After Kogan 1975)

Allelochemic factors	Effect on insect response
A. Allomones	Give adaptive advantage to host plant
Repellent	Orient insect away from plant
Locomotor stimulants	Increase walking or flight activity
Suppressants	Inhibit biting or piercing
Oviposition inhibitors	Interrupt egg-laying
Metabolic inhibitors	Disrupt normal physiological processes
B. Kairomones	Give adaptive advantage to insect pest
Attractants	Orient insect towards plants
Arrestants	Slow or stop movement
Excitants	Evoke biting or piercing
Feeding stimulants	Increase duration and rate of feeding
Oviposition inducers	Induce egg-laying
Metabolic regulators	Regulate normal life processes

Many aspects of plant resistance to insects can be explained in part by the effects of chemicals (Guthrie and Walter 1961). The effects on behavioral (non-preference) and physiological responses (antibiosis) of the insects are presented in Table 3.3. These chemicals are of considerable importance between plants and insects (Whittaker and Feeny 1971) and have been termed allelochemic factors. Insect oviposition occurs when the plant possesses certain oviposition inducers, but is interrupted if oviposition inhibitors are present.

Resistance in plants may not only be affected by the presence of allelochemic factors, but even their mere absence in a host may drastically alter its resistance to one insect pest species or the other (Table 3.3). Cucurbitacins, tetracyclic triterpenoids, are feeding attractants to the spotted cucumber beetle *Diabrotica undecimpunctata howardi*. Chambliss and Jones (1966a,b) demonstrated a quantitative relationship between cucurbitacins and the degree of feeding by these beetles. However, cucurbitacin-free cucumbers were susceptible to the two-spotted spider mite *(Tetranychus urticae)*, while those containing cucurbitacin were mite-resistant (De Costa and Jones 1971). There was a strong correlation between the total Cucurbitacin content in *Cucurbita* species and the extent of aggregation and feeding by both southern corn root worm *(D. undecimpunctata)* and the western corn root worm *(D. virgifera)* (Metcalf et al. 1982). Thus cucurbitacins function ambivalently as plant protectants on one hand, and feeding attractants on the other.

The presence of adult cowpea curculio *(C. aeneus)* on pods of certain lines of southern peas *(Vigna sinensis)* was due to the feeding stimulant present in pods and the success of adults in penetrating the pods accounted for most of the differences between resistant and susceptible lines (Cuthbert and Davis 1972). Twenty six varieties of southern pea were evaluated in the field for resistance to *C. aeneus*, then pods were analyzed in the laboratory for concentration of reducing sugars, starches and total nitrogen. There were significant positive correlations between percent concentration of total carbohydrates in the hull and feeding punctures in the hull and pea. Significant positive correlations occurred between hull, seed nitrogen and feeding punctures (Chalfant and Gaines 1973).

Sinigrin (oil glucoside occurring in various species of Cruciferae) may be a specific stimulus for host selection by the cabbage aphid *(Brevicoryne brassicae)*. The resistance to the aphid can be determined by the waxy or non-waxy nature of *Brassica* plants (K.F. Thompson 1963). To test the effect, sinigrin was introduced into young leaves of the broad bean *(Vicia faba)*, which is not a host of aphid. After 24 h an average of 93% of the aphids were found settled and feeding on sinigrin-treated leaf (Wensler 1962). Munakata and Okamoto (1967) identified Oryzanone (p-methyl acetophenone) as a larval attractant, and benzoic and salicylic acids as larval growth regulators in the rice plant. A higher nitrogen content and a greater percentage of starch have been recorded in certain yellow stem borers *(Tryporyza incertulas)* in susceptible varieties than in resistant ones (Panda et al. 1975). Incorporation of fresh water extract of susceptible and resistant varieties of rice into artificial diets showed that striped borer larvae had a definite preference for the susceptible Rexoro variety over resistant TKM 6 and suffered poor growth in Taitung 16, another resistant variety (Das 1976).

The resistance of *Melilotus infesta* to the sweet clover weevil *(Sitona cylindricollis)* was due to deterrent β, which has been isolated from the leaves (Akeson et al. 1969).

High concentrations of cyanide in sorghum were correlated with a reduction in feeding by grasshoppers and by first instar larvae of *Chilo partellus* and high concentrations of phenolic acids were correlated with reduced feeding by various grasshoppers and by plant hoppers (Woodhead et al. 1980).

Ethanol extracts of the glandular hairs, leaves or fruit of tomato, *Lycopersicon esculentum*, plants were found to contain materials repellent to the adult female immature forms of mites *T. urticae* and *T. cinnabarinus* but not to adult males of these species and to the adult females of *T. turkestani*. The repellent had a half-life of ca. 1.7 h (Cantello et al. 1974).

b) Antibiosis. The term antibiosis is used in its usual sense as the tendency to prevent, injure, or destroy (insect) life. Painter (1941) proposed it for those adverse effects on the insect life history which result when the insect uses a resistant variety of the host plant or species for food. The following adverse physiological metabolic effects of temporary or a permanent nature occurs when the insect feeds on a resistant plant:

– Death of first instar nymphs or larvae, due to which infestation rate in resistant and susceptible plants varies from zero to high, respectively.
– Decline in size and weight of larvae, delayed larval period exposing young larvae longer to predators, reduced fecundity or emergence of short-lived adults, setting a limit to the time available to female insects for mating and egg-laying.
– Continuity of nutrient-poor food reserves affecting the survival ability of hibernation or even aestivating insects.
– Mortality just before the adult stage, which may result in lowering of insect population.
– Frequent restlessness and other behavioral pecularities.

Various combinations of these effects result from feeding on different resistant varieties. Four possible physiological explanations were suggested by Painter (1951):

– The deleterious effects of specific chemicals including toxins;
– food materials present but for some reason not available to the insect;

— the lack of specific food materials in parts of plants eaten; and
— the presence of materials so repellent that the insects will not eat on such resistant plants and virtually starve to death.

Some of the cases of antibiosis are discussed in detail in the following paragraphs and others are presented in Table 3.4.

The ether-insoluble residues of aqueous concentrates contained an additional corn borer inhibitor(s) termed Resistance Factor B (RFB) by Loomis et al. (1957). A "lethal

Table 3.4. Antibiosis as a mechanism of resistance in some of the crop plants

Crop	Insect pests	Reference
Cotton, *Gossypium arboreum*	Cotton boll weevil *(Anthonomus grandis)*	J.A. Bailey et al. (1967)
G. davidsonii	Cotton boll weevil	J.A. Bailey et al. (1967),
G. thurberi	Cotton boll weevil	J.A. Bailey et al. (1967), J.N. Jenkins et al. (1964),
G. barbadense	Cotton boll wevil	J.H. Black and Leigh (1963)
Rice, *Oryza sativa*	Rice green leafhopper *(Nephottetix* spp.)	Athwal and Pathak (1972)
	Brown planthopper *(Nilaparvata lugens)*	Athwal and Pathak (1972)
	White backed planthopper *(Sogatella furcifera)*	Vaidya and Kalode (1981), Gunathilagaraj and Chelliah
	Striped rice borer *(Chilo suppressalis)*	(1985), Athwal and Pathak (1972)
	Gall-midge *(Pachydiplosis oryzae)*	S.V.S. Shastry et al. (1972), M.V.S. Shastry and Prakasa Rao (1973)
Oats, *Avena sativa*	Green bug *(Schizaphis graminum)*	Starks and Burton (1977), R.L. Wilson et al. (1978)
Mungbean, *Vigna radiata*	Weevil *(Callosobruchus chinensis)*	Talekar and Lin (1981)
Alfalfa, *Medicago sativa*	Pea aphid *(Acyrthosiphon pisum)* Weevil *(Hypera postica)*	D.V. Glover and Stanford (1966), Pedersen et al. (1976), T.E. Thompson et al. (1978), Shade et al. (1975)
Sweet clover, *Melilotus* spp.	Aphid *(Acyrthosiphon kondoi)*	Manglitz and Gorz (1968)
Soybean, *Glycine max*	Mexican bean beetle *(Epilachna varivestris)*	Schillinger (1976)
Corn, *Zea mays*	European corn borer *(Ostrinia nubilalis)*	Guthrie (1969)
	Corn ear worm *(Heliothis zea)*	Walter (1962), N.W. Widstrom and Hamm (1969), Douglas and Eckhardt (1957)
	Corn leaf aphid *(Rhopalosiphum maidis)*	Everly (1967)
	Stem borer *(Chilo zonellus)*	V.K. Sharma and Chatterji (1971)
Wheat, *Triticum aestivum*	Hessian fly *(Mayetiola destructor)*	Gallun and Reitz (1971)
	Grain aphid *(Sitobion avenae)*	Lowe (1984)

silk factor" was reported (Walter 1957) in the silks of some lines of corn resistant to the corn ear worm larvae *(Heliothis zea)*. The "lethal silk factor" was confirmed by Knapp et al. (1967) in an experiment using lyophilized silk powder from known resistant lines as a medium for rearing the ear worm larvae. Extracts from the leaves of resistant corn varieties had deterrent substances which tend to mitigate the feeding of the larvae of the European corn borer *(Ostrinia nubilalis)*. These components were ether-soluble and called RFA. RFA contained two distinct compounds, one 6-methoxybenzoxazolinone (6-MBOA) and the other 2, 4-dihydroxy-7 methoxy-1, 4-benzoxazine-3-1 (DIMBOA). Klun and Brindley (1966) showed that 6-MBOA was not an in vivo constituent of corn tissue and the DIMBOA-glucoside is the in vitro precursor of 6-MBOA in corn. Later Klun et al. (1967) observed that when the DIMBOA was isolated from the corn seedlings and bioassayed using an artificial diet for the European corn borer, larval development was inhibited and 25% mortality resulted. The correlation between concentration of DIMBOA in plant tissues and level of resistance to the leaf feeding by the first broad of European corn borer was highly significant for the inbreds and single crosses (Klun et al. 1970). The relation of DIMBOA and resistance to the corn leaf aphid was studied by Long et al. (1976, 1977). The DIMBOA in maize inbreds showed a highly significant correlation with corn leaf aphid infestation.

High concentration of cyanogenetic glycosides phaseolunation and lotaustrin found in the tissues of *Phaseolus* inhibited feeding of Mexican bean beetle, *Epilachna varivestris*, and were toxic, suggesting that bean varieties containing high levels of glycosides might be resistant to the bean beetle (Nayar and Frankel 1963).

Certain pentoses in bean *(Phaseolus vulgaris)* have been shown to inhibit the growth of the weevil *Callosobruchus chinensis* (Friend 1958).

The inadequate sugar supply of plants at the critical stages of development of insect pests may cause resistance. The larvae of European corn borer (ECB), *O. nubilalis*, require glucose till fourth instar and deficiency of sugar until this stage of pest may cause antibiosis (Beck 1957). Scott and Guthrie (1967) related ECB resistance in some corn inbreds to ascorbic acid deficiency. The silks of certain corn lines resistant to ear worm have lower concentrations of amino acids than those of susceptible lines (Knapp 1966). Similar results with regard to amino acids were obtained in peas to pea aphid *A. pisum* (Auclair 1957). He also found that aphids on resistant plants tended to grow more slowly, exuded less honeydew and produced less offspring. The varieties of pea susceptible to pea aphid contained more nitrogen and less sugar than did resistant varieties (Auclair et al. 1957, Maltais and Auclair 1957). Lack of thiamin, nicotinic acid, folic acid, or choline caused a significant decline in the fecundity of *Dacus dorsalis* (Hagen 1958).

Sogawa and Pathak (1970) suggested the lower asparagine content in the Mudgo rice variety as a factor for resistance to brown planthopper *(Nilaparvata lugens)*, biotype 1. The brown planthopper did little feeding on Mudgo, since it contained smaller quantity of asparagine as compared to susceptible varieties. The females exhibited a strong attraction to this amino acid. The low asparagine, therefore, caused restricted feeding and less fecundity.

Another aspect of antibiosis was related to the restlessness factor (Cartier 1963). On resistant alfalfa plants, the spotted alfalfa aphids *(Therioaphis maculata)* became restless in 1 to 4 h, and eventually died or left the plant, little or no honeydew being produced. However, on susceptible plants honeydew was produced profusely. The mortality of aphids on highly resistant plants is caused by starvation or dessication, resulting from failure to ingest a sufficient quantity of plant sap (McMurty and Stanford 1960). Green bugs *(Schizaphis graminum)* moved more on resistant varieties of four species of small grains than on susceptible varieties. Increased mobility could reduce feeding and thus reduce the damage, as well as cause adverse effects on the fecundity of greenbugs (Starks and Burton 1977). Some lines of oats (Table 3.4) *(Avena sativa)* showed antibiosis to biotype C of green bug (R.L. Wilson et al. 1978).

Resistance to the rice stem borers has been linked with the silica content of the rice (Nakano et al. 1961, Sasamoto 1961) and the varietal differences in the total silica content have been recorded (Djamin and Pathak 1968 and Panda et al. 1975). Larvae feeding on rice varieties containing high silica exhibited typical antibiotic effects. This may result in worn-out mandibles (Pathak 1967).

In the case of the red raspberry variety Canby resistant to aphids *(Amphorophora agathonica)*, the aphids exhibited reduced size and fecundity and extensive mortality (Kennedy and Schaefers 1974) and Canby is virtually immune to infestation (Kennedy and Schaefers 1975). The resistance of Canby was related to the levels of both solids (largely sugars) and nitrogenous compounds in the food of *A. agathonica* because the ingestate was more dilute than that obtained from the susceptible variety. Inverse correlation between the susceptibility of tissues to woolly aphids and the ratio of phenolics to α-amino nitrogen in apple tissues was observed (Sen Gupta and Miles 1975).

Cotton plants contain subepidermal (pigments) glands in their above ground parts that vary considerably in density and prominence among species of *Gossypium* and even between races of the same species. In cultivated upland cotton *(G. hirsutum)*, these glands contain about 40% gossypol, a polyphenolic compound. The lines of upland cotton with a high gossypol content were evaluated against the boll worm *(H. zea)* and the tobacco bud worm *(H. virescens)* in the laboratory and in cage tests (Lukefahr and Houghtaling 1969). Line XG-15 (average content of gossypol 1.7% dry weight basis) had significantly reduced populations of tobacco bud worms in a large replicated field cage test. It was utilized less efficiently by the boll worm than a standard line containing 0.5% gossypol (Shaver et al. 1970). The gossypol was equally toxic to the boll worm and tobacco bud worm (Lukefahr and Martin 1966). Less than 30% of the larvae of both species reached the pupal stage when the diet contained 0.2% gossypol. Quercetin inhibited the growth of boll worm larvae, but there was relatively very little mortality at the concentrations below 0.8%. However, this chemical killed more than 70% of bud worm larvae at the 0.1% concentration. Rutin was found to be less toxic to both species than gossypol and quercetin. It was more toxic at the higher dosages to the bud worm than to the boll worm.

The content of oak tree leaf tannins, which inhibit the growth of winter moth larvae, *Operophtera brumata* increases during the summer and may render leaves less suitable for insect growth by further reducing the availability of nitrogen and perhaps also by influencing leaf palatability (Fenny 1970).

High saponin concentration was correlated with resistance to pea aphids, *Macrosyphium pisi* in alfalfa (Pederson et al. 1976).

The absence of some nutritional materials, such as vitamins, vitamin-like substances, sugars or essential amino acids in the particular part of the resistant plant eaten by the insect may also cause antibiosis.

Antibiosis may be involved as a resistance factor influencing larval survival for the sugarcane borer *(Diatraea saccharalis)* in rice (Oliver and Gifford 1975). Varietal resistance to the leafhopper and planthopper is indicated by the higher mortality of nymphs feeding on the resistant variety (Pathak 1969). Resistance to attack in lettuce by root aphid, *(Pemphigus bursarius)*, may result from antibiosis (J.A. Dunn 1960). Clones of alfalfa showed partial antibiosis against pea aphid (Carnahan et al. 1963). Antibiosis operated for resistance in the Washington variety of red raspberry against the aphids *A. agathonica* (Kennedy and Schaefers 1974) and against *Chilo zonellus* in maize (V.K. Sharma and Chatterji 1971). When larvae of the black swallowtail butterfly *(Papilio polyxenes)* were reared on celery leaves *(Apium graveolens)* cultured in solutions of sinigrin, growth and development were substantially reduced. At concentrations of 0.1% (fresh weight of leaf) or higher, sinigrin caused 100% larval mortality (Erickson and Fenny 1974). The exudate from secretory trichomes in *Medicago disciformis* caused antibiosis to alfalfa weevil *(Hypera postica)* larvae. The larvae did not survive when placed on the plant, nor did they survive when exudate is applied to first instars topically. At lower concentrations of exudate, larvae showed a reduced rate of development and a reduced rate of feeding (Shade et al. 1975). A portion of alfalfa *(Medicago sativa)* line PI 247790 had convulsed, dehydrated and caused larval death before pupation. Another mechanism characterized by delayed larval development was also observed in this line (T.E. Thompson et al. 1978).

c) **Tolerance**. Tolerance is the mechanism of resistance in which a plant shows an ability to grow and reproduce itself or to repair injury to a marked degree in spite of supporting a population approximately equal to that damaging a susceptible host. Resistant plants may be tolerant, surviving under levels of infestation that would kill or severely injure susceptible plants (Painter 1951). Resistant hybrids of corn supported levels of corn ear worm infestation that seriously damaged susceptible hybrids (Wiseman et al. 1972). Mudgo and DV 139 rice cultivars had high larval survival and growth as well as high field borer populations, but suffered less damage because of their tolerance (Das 1976).

Tolerance as a mechanism has also been reported against greenbug in wheat (Curtis et al. 1960 and Porter and Daniels 1963); in barley (O.D. Smith et al. 1962); in oats (Gardenhire 1964); in sorghum (Weibel et al. 1972); and in rye (Livers and Harvey 1969) and to western corn root worm in corn (Sifuentes and Painter 1964).

The tolerance differs from non-preference and antibiosis in its mechanism. Non-preference and antibiosis require an active insect response or lack of response. However, tolerance response requires a thorough understanding of the ways in which plants are injured by insects, as well as the ways in which plants may repair the damage. Tolerance is more subject to variation as a result of environmental conditions than non-preference and antibiosis. The age or size of the plant and the size of insect population strongly influence the degree of tolerance. The general vigour of a plant greatly affect its tolerance to insect attack.

3.4 The Combination of Factors

Of these mechanisms of resistance discussed, one or a combination of any of the three is present in most cases of resistance. The three mechanisms are inter-related. Such a basic triad of resistance relationships has usually been found to result from independent genetic characters, which are, however, interrelated in their effects.

The leaf sheaths of certain wheat plants and one variety of oats that were resistant to hessian fly attack appeared to have a much more complete and even distribution of silica deposits on their surface than do those of susceptible varieties. Silica depositions on wheat plant tissue may be one of the factors related to resistance to hessian fly, other factors may also be involved (Miller et al. 1960). It has been reported that wheats with highly pubescent leaves were largely avoided for oviposition by cereal leaf beetle *(Oulema melanopus)* (Leisle 1974). Schillinger and Gallun (1968) observed that density of pubescence in four wheat lines adversely affected oviposition, viability of eggs and growth and survival of larvae.

The resistance in rice to stem borers and to leafhopper and planthopper appears to be based on non-preference, antibiosis or both (Pathak 1969). Rice resistance to stem borer *(Chilo suppressalis)* results from the interaction of leaf blades with a hairy upper surface, tight leaf sheath wrapping, small stems with ridged surface and thicker hypo-dermal layers (Patanakamjorn and Pathak 1967). Resistant varieties of rice generally were not preferred to green leaf hopper *(Nephotettix virescens)*. Also insects caged on these plants suffered high mortality. On highly resistant varieties only 0–3% of the first instars reached the adult stage, whereas on susceptible varieties 76–90% became adults (Cheng and Pathak 1972). Varietal resistance to brown planthopper *(Nilaparvata lugens)* in rice was studied by Seetharaman et al. (1984). The resistant cultivars showed less adult preference for shelter, oviposition and feeding. Survival of nymphs and progeny developed were also adversely affected.

Resistance in corn to the corn ear worm *(H. zea)* has been ascribed to long husks, tight husks, blunt ear tips, flinty tip kernels, silk balling and starchness of kernels (Luckmann et al. 1964 and Wiseman et al. 1972). Multiple factor resistance in maize to European corn borer (ECB) was reported by Rojanaridpiched et al. (1964). Resistance to second generation ECB was significantly correlated with the silica content in the sheath and collar tissues. Additionally, relatively high DIMBOA content was found in the leaf sheath and collar tissues of some lines, and DIMBOA was shown to have a secondary role in second-generation resistance of some lines. Resistance of the first generation ECB was highly correlated with DIMBOA content in the whorl tissues, but silica and lignin appeared to contribute to this generation resistance as well.

Resistance in cotton to the pink boll worm *(Pectinophora gossypiella)* involves absence of bracts, glabrous leaves, cell proliferation, high gossypol content, and nectar-less character (Agrawal et al. 1976). Also in cotton, frego bract, red plant colour, in-creased pubescence and rapid fruit set provide resistance to the boll weevil *(Anthono-mus grandis)* and smooth-leaved, nectar-less and high gossypol plants offer resistance to *Heliothis* spp. (Lukefahr et al. 1966).

A variety which is not preferred does not require the antibiosis or tolerance that must be present in a preferred variety of the same level of resistance. A plant that exhibits high tolerance and low antibiosis towards a given insect may show the same level of resistance as one that has a high degree of antibiosis and an average degree of tolerance. The gene(s) for one or more types of resistance may be recombined. This type of recombination in addition to the possible multiple genetic factors in the plant may be the basis of any one of the three characteristics. The genetic expression of these three characters of resistance is frequently modified by various ecological conditions and by other genes.

3.5 The Biotypes of Insect Pest

The development of insect biotypes capable of surviving on resistant plants limits the use of resistant varieties. Biotypes in insect pests are comparatively infrequent because of the insect's own complex physiology and the pest resistance in host plants being often related to the host-finding behaviour of insects. The biotypes are distinguished on the basis of their interaction with differential host varieties or clones of the host plants. The biotypes are not necessarily similar to the geographical races or insect populations which may be distinguishable on other biological grounds. The biotypes may occupy definite geographical areas, as has been reported (Gallun 1965) for those of hessian fly *(M. destructor)*.

The development of biotypes in insect pests is obviously traced to the severe selection pressure exerted by the resistant crop variety. When antibiosis is the component of resistance, such selection pressure is most effective if it results in mass mortality (Panda 1979). Gallun (1972) observed that tolerance or non-preference is the main component of resistance inherent in the plant.

Ten biotypes of hessian fly, designated Great Plains, A to G and J,L have been identified in the field or selected in the laboratory (Gallun 1977, Sosa 1978, 1981). These biotypes differ in ability to cause stunting and to survive on wheats having different genes for resistance. The mechanism of resistance is antibiosis, i.e. first instar larvae of avirulent biotypes die after feeding on resistant plants.

Pathak (1975) noted that insect biotypes are more likely to develop on plants which have monogenic resistance than on plants with polygenic resistance. Thus, the genetic base of at least major crop plants should be broadened to include pest-resistant characteristics to equip crops with inherent protection.

In a few cases, biotypes capable of surviving on resistant plants may be larger and more vigorous; insects such as the pea aphid, but generally biotypes, are adapted to the effects of specific genes for resistance. A knowledge of insect biotypes can help plant breeders to identify diverse sources of resistance and to breed for multigenic resistance.

A list of some of the known biotypes is presented in Table 3.5. Six of the 11 insects with known biotypes are aphids, which could be due to parthenogenetic reproduction and relatively a short life cycle. An aphid that can survive on a resistant plant can build up a new biotype within one or two crop seasons.

Table 3.5. A list of known insect pest biotypes

Crop	Insect	Biotypes	Reference
Wheat	Hessian fly *(Mayetiola destructor)*	GP, A to G(8) J and L (2)	Gallun and Hatchett (1968) Sosa (1978, 1981)
	Greenbug *(Schizaphis graminum)*	4	Wood et al. (1969), Boozaya-Angoon et al. (1981), Porter et al. (1982)
Rice	Brown planthopper *(Nilaparvata lugens)*	4	IRRI (1974)
	[a]Gall midge *(Pachydiplosis oryzae)*	–	S.V.S. Shastry et al. (1972), Kush (1977a)
	[a]Green leafhopper *(Nephotettix* spp.)	–	
Raspberry	Aphid *(Amphorophora rubi)*	4	J.B. Briggs (1959, 1965), R.L. Knight et al. (1960), Keep and R.L. Knight (1967)
Pea alfalfa	Pea aphid *(Macrosyphium pisi)*	3–9	Harrington (1943), Cartier et al. (1965)
Alfalfa	Spotted alfalfa aphid *(Therioaphis maculata)*	6	M.V. Nielson et al. (1970)
Sorghum, corn	Corn leaf aphid *(Rhopalosiphum maidis)*	4	Painter and Pathak (1962)
Apple	Woolly aphid *(Eriosoma lanigerum)*	3	Sen Gupta (1969)
Beans	Bean aphid *(Aphis fabae)*	2	Pathak (1970)

[a] Not established.

3.6 Genetics of Virulence

The first report on the inheritance of virulence were with hessian fly. Its genetics is complicated by the fact that when male flies form sperm, the parental haploid genome (n = 8) is eliminated and only the maternal chromosomes are transmitted (Gallun and Hatchett 1969). As a result, F_2 and back-cross genotypes from different races vary according to the direction of the original cross between the parents. F_1 flies used as males breed as though homozygous because they transmit on maternal chromosomes. The F_1 females breed as heterozygotes, showing normal transmission of both genomes subject to reassortment and crossing-over at meiosis.

Hatchett and Gallun (1970) studied crosses between three races of hessian fly, distinguished by their reactions (Table 3.6). Monon wheat carries the single dominant gene (H_3) for resistance (R.M. Caldwell et al. 1946), whereas the resistance of Seneca is undetermined. Race GP is avirulent on all wheats carrying oligogenic race-specific resistance. Hatchett and Gallun (1970) crossed GP and E races, their F_1, F_2 and back-cross progenies were tested on seedlings of Monon. They found that avirulence was dominant to virulence. Virulence was governed by a single gene. Similar results were obtained on Seneca.

Table 3.6. Differential reactions of three races of hessian fly *(M. destructor)*. S = high or susceptible; R = low or resistant. (Hatchett 1969)

Wheat variety	Genotype	Hessian fly races		
		GP	A	E
Turkey	h	S	S	S
Seneca	?	R	S	R
Monon	H_3	R	R	S

Genetic studies of virulence of biotypes B and C of hessian fly suggested that (1) the gene for virulence to Monon wheat in biotype B flies and the gene for virulence in C biotype to attack Knox 62 are on separate chromosomes; (2) the allele M in biotype C that causes it to be avirulent to Monon wheat is located near the centromere, the allele K in biotype B that causes it to be avirulent to Knox 62 is located at a considerable distance from the centromere; (3) during spermatogenesis in the hessian fly, paternally derived centromeres and the attached chromatin are eliminated; and (4) crossing-over occurs in the male hessian fly prior to elimination of the chromatin (Gallun 1978).

Resistance of rice varieties to brown planthopper *(N. lugens)* is based on major genes; it was assumed that a gene-for-gene relationship between host and pest may exist. However, the mode of inheritance of virulence and the response of the biotypes to selection, together with the previously reported wide variation within each biotype and large overlap between them in virulence, is all consistent with polygenic determination of virulence (Den Hollander and Pathak 1981).

3.7 Genetics of Resistance

Plants can be evaluated in the field if there is high natural insect population or in the greenhouse with field-collected or laboratory-reared insects. However, the field culture of insects may have variability for virulence, so that it is better if laboratory-reared insects are used for screening.

Resistance may be determined based on plant injury or symptoms of insect attack or on the reaction of the insects to the plant. The insect damage may be in the form of defoliated plants as in the case of striped cucumber beetle on a squash (Nath and Hall 1963), stunted plants, as with hessian fly on wheat (Gallun 1965), or death of the plant, as in hopperburn "caused by brown plant hopper in rice" (Athwal et al. 1971). Insect reactions to resistant plants may be death (hessian fly and wheat stem sawfly), reduced fecundity (spotted alfalfa aphid and greenbug), loss of weight (cereal leaf beetle), restless-ness (aphids), a longer period between stages in the life cycle of the insect, or avoidance of the plant for feeding and/or ovipositing.

The pest-host interactions are scored on a scale based on the reaction of the plant to the insect and, less often, by the reaction of the insect to the plant. The reactions of parents, F_1, F_2 and F_3 progenies are used to study the dominant/recessive or incompletely dominant and quantitative or qualitative nature of inheritance.

In the subsequent discussions, the genetics of resistance to some of the insect pests where systematic studies have been conducted will be presented in detail and other reports are presented in Table 3.7. It is apparent from Table 3.7 that the studies on inheritance of resistance have been conducted in almost all the important cereals, legumes, vegetables and fruits against their major insect pests.

Table 3.7. Some of the reports on the inheritance of resistance to insect pests in the important crop species

Crop species	Insect pest	Inheritance	Reference
Wheat	Green bug (*Schizaphis graminum*)	A single recessive gene was responsible for resistance	Painter and Peters (1956), Daniels and Porter (1958), Curtis et al. (1960), Porter and Daniels (1963)
	Cereal leaf beetle (*Oulema melanopus*)	Resistance was due to pubescence and was inherited quantitatively. The additive gene action was more important	Ringlund and Everson (1968)
	Wheat stem sawfly (*Cephus cinctus*)	Resistance was due to solid stem and at least three genes that had equal effect were responsible	McKenzie (1965)
	Rice weevil (*Sitophilus oryzae*)	Resistance is polygenic in nature	V.S. Singh and Bhatia (1979)
Barley	Hessian fly (*Mayetiola destructor*)	A single dominant gene Hf at cooler temperature and complementary gene action at higher temperature was expressed	Olembo et al. (1966)
	Corn leaf aphid (*Rhopalosiphum maidis*)	1. Resistance was conditioned by a single recessive gene	Dayani and Bakshi (1978b)
		2. Resistance was dominant and controlled by one pair of genes	Narula et al. (1984)
	Cereal leaf beetle (*O. melanopus*)	A recessive gene for tolerance was reported in barley variety C 166	Hahn (1968)
	Green bug (*S. graminum*)	A single dominant gene Grb was responsible for resistance	Gardenhire and Chada (1961), Gardenhire (1965), O.D. Smith et al. (1962)
Rye	Green bug (*S. graminum*)	Tolerance is inherited as monogenic dominant	Livers and Harvey (1969)
Rice	White backed planthopper (*Sogatella furcifera*)	1. A single dominant or recessive gene was reported for resistance	Sidhu et al. (1979), Nair et al. (1982), Jayraj and Murty (1983)

Table 3.7 (continued)

Crop species	Insect pest	Inheritance	Reference
		2. Resistance is also governed by two independent dominant genes or by one dominant and one recessive gene	Angles et al. (1981)
	Rice gall midge *(Pachydiplosis oryzae)*	1. One dominant gene for resistance was postulated in W 1263	Satyanarayanaiah and Reddy (1972)
		2. Two and four genes for resistance were postulated in W 1263 and Ptb 18, respectively	Shastry et al. (1972)
		3. Three recessive genes for resistance were reported in W 1263 and W 12708	M.V.S. Shastry and Prakasa Rao (1973)
	Striped borer *(Chilo suppressalis)*	1. Using borer infestation as a criterion the field resistance of Giza 14 was observed to be under polygenic control	Koshiary et al. (1957)
		2. Using white head as criterion field resistance of TKM 6 was simply inherited	Athwal and Pathak (1972)
	Yellow borer *(Tryporyza incertulas)*	Resistance is governed by a single recessive gene at heading stage	Dutt et al. (1980)
Sorghum	Green bug *(S. graminum)*	A single incompletely dominant gene was responsible for resistance	Weibel et al. (1972), Starks et al. (1972a, 1972b)
	Shoot fly *(Atherigona varia soccata)*	1. Resistance was observed to be quantitative	Starks et al. (1970), Halalli et al. (1982)
		2. Heritability for shootfly resistance was around 15 to 25% under high to moderate levels of infestation	Harwood et al. (1973), Borikar and Chopde (1980)
	Chinch bug *(Blissus leucopterus)*	A single dominant gene was postulated for resistance	Painter (1951)
Corn	European corn borer *(Ostrinia nubilalis)*	1. A single recessive/dominant gene was responsible for resistance	Marston (1930), Penny and Dicke (1957)
		2. Two and three genes were responsible for resistance	R. Singh (1953), Penny and Dicke (1956)
		3. Resistance was due to cummulative effect of several genes	Schlosberg and Baker (1948), Scott et al. (1964, 1965), Scott and Guthrie (1967), M.S. Chiang and Hudson (1973)

Table 3.7 (continued)

Crop species	Insect pest	Inheritance	Reference
		4. Additive gene action was more important Genes for resistance to 1st and II brood may be different but some of the genes may contribute to both broods	C.W. Jennings et al. (1974)
	Corn ear worm (Heliothis zea)	1. Resistance is inherited quantitatively	N.W. Widstrom and Hamm (1969), Keaster et al. (1972)
		2. The additive effect was more important in the cross of sweet corn and dominance effect was more important in dent corn cross	N.W. Widstrom and McMillian (1973)
	Western corn root worm (Diabrotica vergifera)	A single recessive gene was reported for resistance	Sifuentes and Painter (1964)
	Fall army worm (Spodoptera frugiperda)	The mode of inheritance of resistance was quantitative	N.B. Widstrom et al. (1972)
Pear Millet	Chinch bug (B. leucopterus)	Tolerance was due to a single dominant gene	Starks et al. (1982)
Oats	Green bug (S. graminum)	1. Tolerance was inherited as monogenic	Gardenhire (1964)
		2. Resistance was due to single dominant gene. The gene that controlled resistance to one biotype was not effective against other biotype	Boozaya-Angoon et al. (1981)
Cotton	Thrips (Thrips spp.)	Pilosity influences resistance. Three dominant genes affected pilosity (nonpreference)	Ramey (1962)
Snap bean	Leafhopper (Empoasca kraemeri)	Resistance was recessive and quantitative in nature	Lyman and Cardona (1982)
Legume and forage	Sweet clover aphid (Therioaphis richmi)	A single dominant gene conveyed resistance. An additional complementary gene appeared to be present in some resistant clovers	Manglitz and Gorz (1968)
	The pea aphid of alfalfa (Macrosyphium pisi)	One dominant gene was found to confer resistance	D.V. Glover and Stanford (1966)
	Spotted alfalfa aphid (T. maculata)	1. Resistance was observed to be quantitative	C. Glover and Melton (1966)
		2. Resistance could be both vertical and/or horizontal	M.W. Nielson and Kuehl (1982), M.W. Nielson and Olson (1982)

Table 3.7 (continued)

Crop species	Insect pest	Inheritance	Reference
Soybean	Cyst nematode *(Heterodera glycines)*	Resistance is the result of three independently inherited recessive genes	B.E. Caldwell et al. (1960)
Cowpea	Weevil *(Callosobruchus maculatus)*	Resistance to cowpea weevils has additive dominance and maternal components	Fatunla and Badaru (1983)
Vegetables	Lettuce root aphid *(Pemphigus bursarius)*	Resistance was controlled by extranuclear factor. Modifying genes might also be involved	J.A. Dunn (1974)
	Lettuce leaf aphid *(Nasonovia ribis nigri)*	Resistance was due to one dominant gene	Eenink et al. (1982a,b)
	The melon aphid *(Aphis gossypii)*	A single dominant gene, Ag was identified for resistance	Kishaba et al. (1971), Bohn et al. (1973)
	Striped cucumber beetle *(Acalymma vittatum)*	Resistance was governed by several genes, and additive gene action was more important	Nath and Hall (1963)
	The squash bug *(Anasa tristis)*	Resistance was observed to be quantitative, and additive variance was higher than dominance variance	Benepol and Hall (1967)
	Red pumpkin beetle *(Aulacophora foveicollis)*	A single dominant gene, Af conferred the resistance	Vashistha and Choudhary (1974), Chambliss and Jones (1966a)
	Fruit fly *(Dacus cucurbitae)*	The resistance in watermelon was controlled by a single dominant gene, Fwr. Similar inheritance was reported in *Cucurbita maxima* and Fr gene symbol was proposed	Nath et al. (1976b), Khandelwal and Nath (1978)
	Green peach aphid *(Myzus persicae)*	Resistance appeared to be partially dominant in tuber bearing *Solanum*, and genotype-environment interaction was low	Sams et al. (1976)
Trees and fruits	Rosy leaf curling aphid of apple *(Seppaphis devecta)*	A single dominant gene was responsible, which was designated as Sd	Alston and Briggs (1968)
	Rosy apple aphid *(Dysaphis phantaginea)*	A single dominant gene designated as Sm_h was reported	Alston and Briggs (1970)
	The wooly apple aphid *(Eriosoma lanigerum)*	A single dominant gene Er was responsible for resistance	Painter (1951)
	Aphid *(Amphorophora rubi)*	Immunity is governed by a single dominant gene linked to a semilethal gene	Daubeny (1966)

3.7.1 Wheat

a) Hessian fly *(M. destructor)*. Twelve genes for resistance to this insect have been identified; ten are dominant and one is recessive (Gallun and Khush 1979, Stebbins et al. 1980, 1982, 1983, Oettermann et al. 1983). Cartwright and Weibe (1936) and Noble and Suneson (1943) identified Dawson gene, H_1 and H_2; R.M. Caldwell et al. (1946) and Abdel-Malek et al. (1966) identified W 38 gene, H3. The recessive Java gene, H4 was identified by Suneson and Noble (1950). The Ribeiro gene, H5 was described by Shands and Cartwright (1953) and H6 gene in durum, PI 94587 (R.M. Caldwell et al. 1966) was described by Allan et al. (1959). H7 and H8 were identified in Seneca by Patterson and Gallun (1973). Elva gene H9, H10 *(T. turgidum)* was reported in Purdue selection 822–34 to biotype D by Stebbins et al. (1980, 1982). H9 was transferred to common wheat Ella. H9 and H10 were transferred as a gene block from Elva wheat to common wheat selection Stela and 817–2. H10 was separated from H9 using a test cross of selection Stella. Ella (H9H9) was crossed to resistant homozygous lines derived from the Stella test cross to identify lines with H9H9 or H10H10. In a similar manner Ella was crossed to 817–1 selections to identify lines from 817–1 that were either H9H9 or H10H10. In F_3 lines repulsion phase analysis, H9 or H10 were found to be independent (Stebbins et al. 1982). PI 94587 has one dominant gene in addition to the H6 gene and it was designated as H11.

Oettermann et al. (1983) reported a single dominant gene (H12) in Luso cultivar of common wheat to biotype B and D.

The gene-for-gene concept similar to the one observed by Flor (1942, 1955) in *Linum-Melampsora lini* system is reported in *Triticum-Mayetiola destructor* by Hatchett and Gallun (1970).

3.7.2 Rice

a) Brown Planthopper *(N. lugens)*. A night level of resistance to BPH was first discovered in the rice variety Mudgo at IRRI (Pathak et al. 1969). The resistance gene of IR 747B2–6 to BPH was allelic to the dominant gene of Mudgo (Bph 1) (Athwal et al. 1971, Chen and W.L. Chang 1971). The resistance of IR 1154–243 and of IR4–93 was governed by a recessive gene which was also allelic to the resistance gene bph 2 of ASD7 (Athwal et al. 1971, Athwal and Pathak 1972). IR4–93 inherited its resistance from H105, but both the parents of IR 747B2–6 and of IR 1154–243 were susceptible. TKM 6, one of the parents of IR 747B2–6, is homozygous for Bph 1 but is also homozygous for a gene, I–Bph 1 which inhibits Bph 1. Zenith one of the parents of IR 1154–243 may also have a similar inhibitor gene (Martinez and Khush 1974). Resistance in MTU 9, Murunga–137 and Sudurvi–306 was each controlled by a single dominant gene, while H5, Dikwee, and Kashsiung Sen-Yu 12 appeared to have a recessive gene. Crosses of IR9–60 with H5 and Kaohsiung Sen-Yu12 suggested that resistance of these cultivars was due to the same gene (W.L. Chang and Chen 1971) or closely linked genes and the recessive resistance gene of IR9–60, H5, Kaohsiung Sen-Yu12 were identified as bph2 of ASD7 (W.L. Chang 1975). Single dominant genes that are allelic to Bph 1 condition the resistance in Balamawee, Co10, Heenukkulama, MTU-9,

Sinnakayam, SLO 12, Sudhubalawee, Sudurvi 305 and Tibiriwewa. Single recessive genes that are allelic to bph2 govern resistance in the cultivars Anbaw C7, ASD 9, Dikwee 328, Hathiel, Kosatawee, Madayal, Mahadikwee, Malkora, M.I.329, Murunga-kayan 302, Overkaruppan, Palasithari 601, PK-1, Seruvellai, Sinnakaruppan and Vel-lailangayan. A single dominant gene also conveys resistance in Rathu Heenati but it segregates independently of Bph 1 and is designated as Bph 3. Similarly, a single reces-sive gene conveys resistance in Babawee, but it segregates independently of bph2 and is designated bph4. The resistance in Ptb 21 is controlled by one recessive and one dominant gene (Lakshminarayana and Khush 1977). However, a pair of recessive genes were controlling resistance to bph in RP 31-49-2 (susceptible) and Leb Muey (mod-erately resistant) cross (Prasada Rao et al. 1977). Bph 1 and bph2 are closely linked (Athwal et al. 1971) and Bph3 and bph4 are either allelic or closely linked (Sidhu and Khush 1979).

b) Green Leafhopper *(Nephotettix virescens)*. The inheritance of resistance to the green leafhopper (GLH) was studied in 13 cultivars of rice (Siwi and Khush 1977). A single dominant gene that is allelic to the dominant gene Glh 1, originally identified in Pankhari 203 (Athwal et al. 1971) conveys resistance to GLH in Jhingasali. A single dominant gene was located at Glh2 locus in Godalki, Lien-tsan 50, and Palasithari 601 which was originally identified in ASD7 (Athwal et al. 1971). The resistance in Betong, DNJ 97 and H5 is conditioned by dominant genes that are allelic to Glh3 originally found in IR8 (Athwal et al. 1971). In ASD8 (Thuyamalli), ARC 6602, DM 77, DS 1 and Khama 49/8, a single dominant gene Glh5, was identified, which is independent to Glh1, Glh2 and Glh3 and conveys resistance to GLH. A single recessive gene glh4 that is independent of Glh1, Glh2 and Glh3 and conveys resistance to glh in Ptb8 (Chuvan-nari Thavalakkannam). The allelic relationship of glh4 and glh5 is not known.

3.7.3 Raspberry *(Rubus idaeus)*

a) Aphid *(Amphorophora rubi)*. Most of the work in the *Rubus* spp. has been with rubis aphid. The winged forms of this aphid spread several viral diseases.

Four strains (1, 2, 3 and 4) were recognized (J.B. Briggs 1959, R.L. Knight et al. 1960, Keep and R.L. Knight 1967). The relationships between the genes for resistance and the strains of aphid is presented in Table 3.8. Sources of resistance of the ten genes are given in Table 3.9. The combination of A_1 and A_2 or A_1 and A_3 can control *Rubus* aphid strains 1, 2, and 3 (R.L. Knight et al. 1960). Resistance of strain 4 will require A_7 or A_8 (J.B. Briggs 1965). The gene A_{10} can be used for resistance to strain 1, 2, and 3 (Keep and R.L. Knight 1967).

3.7.4 Tomato *(Lycopersicon esculentum)*

a) Nematodes. The inheritance of resistance to the mild strain isolate H of *Coryne-bacterium michiganense* appears to be determined by four genes in tomato. A com-bination of recessive gene (a) and three dominant genes (B, C and D) controls resistance. Three alleles of genes C and D are involved. Incomplete dominance of gene A is ex-

Table 3.8. Relationship between the various resistance genes in *Rubus idaeus* and four strains of *Rubus* aphid. (Maxwell et al. 1972)

Gene in plant	Reaction to aphid strains			
	1	2	3	4
A_1	R	S	R	–
A_2	S	R	S	–
$A_3 A_4$	S	R	S	–
A_5	R	S	S	–
A_6	R	S	S	–
A_7	R	S	S	–
$A_1 + A_2$	R	R	R	–
$A_1 + A_3$	R	R	R	–
$A_1 + A_4$	R	S	R	–
A_8	R	R	R	S
A_{10}	R	R	R	–

Table 3.9. The sources of the ten genes for resistance. (Maxwell et al. 1972)

Resistance gene	Structure
A_1	Baumforth cultivar
A_2	Chief cultivar
A_3	Chief cultivar
A_4	Chief cultivar
A_5	Chief cultivar
A_6	Chief cultivar
A_7	Chief cultivar
A_8	*R. idaeus* var. *strigosum*
A_9	*R. idaeus* var. *strigosum*
A_{10}	*R. occidentalis*

pressed in plants bearing the Aa genotype only in the presence of genes C^2 or D^2. Resistance to the less virulent isolates appears to be determined by the B, C or D genes (Jan De and Honma 1976). The gene-for-gene relationship in *Solanum-Heterodera rostochiensis* system is observed by D.F. Jones and Parrott (1965).

3.8 Location of Gene(s) on Chromosomes

Once the resistance genes are identified in a particular cultivar, it is crossed with other resistant cultivars to study the allelic relationship of res-genes (Lakshminarayana and Khush 1977). If two or more resistant genes are known, their linkage relationships are determined by monosomic analysis, as for hessian fly in wheat (Gallun and Patterson 1977); by trisomic analysis, as for green bug resistance in barley (Gardenhire et al.

1973); by translocation stocks, as for ear worm resistance in maize (Robertson and Walter 1963); or by substitution, as for cereal leaf beetle (D.H. Smith and Webster 1973). After the genes are assigned to chromosomes, they are mapped on the respective linkage groups.

Some of these technique(s) are described here in detail with the help of appropriate examples.

Monosomic analysis in wheat showed that chromosome 5A carries resistance to races A and B of hessian fly *(M. destructor)* and that resistance is controlled by a single dominant gene (Gallun and Patterson 1977). Knox 62 and Purdue 4835 A4–6 hexaploid wheat cultivars possessed H6H6, a dominant gene derived from PI 94587 durum wheat for resistance to race B which was linked with the H3 gene for resistance to race C with a genetic recombination value of about 0.09 (Patterson and Gallun 1977).

A variation of the gene-marked translocation technique, as described by E.G. Anderson (1956), was used by Robertson and Walter (1963) in an attempt to identify chromosomes carrying genes for resistance to ear worm *(Heliothis zea)* injury in two sweet corn inbreds. N.B. Widstrom and Wiseman (1973) tested six sweet corn inbreds for the location of major genes for resistance to ear damage by the corn ear worm. They used a series of waxy marked chromosome 9 translocations. The data indicate that genes for resistance are located on chromosomes 4 and 5 of inbred 245, on chromosome 3 of inbred 20, and on chromosome 8 of inbred La2W. Other chromosomes (4 and 8 for inbred 20, 1 and 3 for inbred 81–1, and 6 for inbred 322) are implicated less strongly in conferring resistance to ear worm by chisquare values at the 10% level of probability.

Chromosome interchanges were used to determine which chromosome arms of resistant maize in B52 contain genes for resistance to sheath collar feeding by second-generation *Ostrinia nubilalis*. Resistant inbred B52 contains a gene or genes for resistance on the long arms of chromosomes 1, 2, 4, and 8 and on the short arms of chromosomes 1, 3, and 5 (Onukogu et al. 1978).

Trisomic analysis was conducted by Ikeda and Kaneda (1982) to locate the Bph3 and bph4 genes for brown planthopper *(N. lugens)* resistance in rice. The Bph3 and bph4 are located on chromosome 7.

3.9 Problems Associated with Breeding for Resistance to Insects

An intimate knowledge of the biology and feeding habits of the insect is essential. Often the knowledge is inadequate, necessitating detailed studies before launching the programme. Insect variability is a problem and this must be closely monitored. Biotypic differences within insect species may cause behaviour different in laboratory-reared colonies than in the field-selected populations. Natural variability in growth, fecundity and other parameters of insect development must be carefully determined and accounted for in tests for resistance. A population of insects must be built up using field or laboratory cultures rather than destroyed. Methods must be designed to examine large numbers of plants for insect infestation or damage, frequently in a very short period of time. Seedling screening techniques for mass selection should be developed. Careful consideration must be given to correlation of resistance in the seedling stage to resistance in the mature plants. The success in developing synthetic varieties of alfalfa in a short period of time to the spotted alfalfa aphid and pea aphid can be attributed largely to seedling rating techniques coupled with simple recurrent selection (Maxwell et al. 1972).

If the chemical nature of resistance can be ascertained early in the programme, a precise chemical testing procedure could be very useful at this stage. Often the resistance is found in the wild relatives, which causes problems in incorporation of resistance and selection of resistant segregants coupled with superior agronomic traits.

A variety developed for resistance to one pest may be susceptible to another insect previously unimportant. Excellent examples include the glanded and glandless condition (gossypol glands) in cotton. Removal of glands through breeding caused greater susceptibility to boll worms (*H. zea* and *H. virescens*) (Lukefahr et al. 1966, Oliver et al. 1970, 1971). In addition, a number of leaf-feeding insects not previously known to cause damage to cotton became serious pests (Maxwell et al. 1965). The frego character in cotton imparts high field resistance to the boll weevil (J.N. Jenkins et al. 1969), but unfortunately causes increased susceptibility to the tarnished plant bug *Lygus lineolaris* (Lincoln et al. 1971). The glabrous condition in cotton which contributes resistance to boll worms causes greater susceptibility to thrips, leaf hoppers, aphids, fleahoppers and certain other pests. Glabrous cottons had the highest bug and leaf hopper populations and the lowest yield (J.C. Bailey 1982).

Leaf pubescence is incorporated into wheats to prevent damage by cereal leaf beetle *(O. melanopus)* (J.J. Roberts et al. 1976, D.H. Smith et al. 1978) and by hessian fly. The pubescent wheats, however, are more heavily infested by air-born wheat curl mite, *Eriophyes tulipae*. The wheat curl mite is the only known vector of wheat streak mosaic virus (WSMV), Slykhuis (1955). Therefore, the incidence of WSMV is higher on pubescent than on glabrous wheats. Both plant hair length and density were directly associated with mite infestation.

4 Genetics of Host-Parasite Interaction

The host and the parasite are the two biological entities involved in the host-parasite interrelationship. The studies on the genetics of resistance in the host and pathogenicity in the pathogen are important to understand the basic aspects of the interrelationships of host and parasite. Depending on the type of interaction, the breeding/management strategies of resistance described in Chap. 6 could be framed.

In this chapter therefore, the genetics of the resistance and virulence and the interaction of host and parasite i.e., the Flor's hypothesis of gene-for-gene relationship and its extension by Vanderplank (i.e., the secondary gene-for-gene hypothesis along with chemical basis of resistance and pathogenesis) are described in detail.

4.1 Genetics of Resistance

The rediscovery of Mendel's laws of inheritance in 1900 provided the foundation necessary for the analysis of the differential reaction of varieties to diseases. The first reported genetic study of resistance to a disease was published in 1905 by Biffen in England. He obtained a 3 (susceptible): 1 (resistant) ratio in the F_2 populations of crosses between resistant wheat variety (Rivet) to yellow rust *(Puccinia striiformis)* and susceptible varieties (Michigan, Bronze and Red King). The F_3 families appeared in the ratio of 1/4 true-breeding susceptible lines, 1/2 segregating lines, and 1/4 true-breeding resistant lines. He also suggested that resistance to ergot was governed by two factor pairs.

The conclusions of Biffen were criticized due to the observations that varieties resistant in certain localities succumbed in others and that resistance varied widely in different seasons. Ward (1902) working with brome-grass species had noted considerable variations in reactions to brown rust. By growing a given strain of rust on a moderately resistant brome for several vegetative generations, he found that pathogenicity could be altered to the point where the rust would attack previously resistant brome species. To explain this, Ward proposed the bridging-host hypothesis, which assumed that the pathogen is plastic and that it can be influenced by the host substrate in ways that change its pathogenicity. Although Biffen (1912) countered with observations that the resistance of Rivet wheat to yellow rust had remained constant over many years, the bridging-host hypothesis undoubtedly discouraged many from attempting to breed for disease resistance. From 1912 to 1970 numerous papers on the inheritance of resistance were published and most of these confirmed the findings of Biffen.

Person and Sidhu (1971) reviewed the work carried out on the inheritance of resistance since Biffen's time. They reviewed more than 1000 published papers almost exclusively concerned with economically important plant species and their fungal, non-fungal parasites including nematodes, bacteria and viruses. They concluded that: (1) Regardless of the species that was involved in the host-parasite interaction, resistance generally segregated in the Mendelian ratios. Resistance was usually found to be determined by "major" rather than by "minor" genes. The alleles for resistance were dominant over those for susceptibility. (2) In a relatively small number of studies the two factor genetic interactions were found. (3) In a relatively small number of studies evidence has been found for linkage of resistant genes. (4) Alternate resistance alleles were distinguished according to specific biotypes.

The information published after 1971 on the mode of inheritance of resistance of diseases in some of the important crop plants against the important plant pathogens is compiled. The literature cited here is by no means a review of the published information on the inheritance of resistance. However, it can be inferred that the mode of inheritance of resistance could be monogenic (Table 4.1), oligogenic (Table 4.2) and

Table 4.1. The monogenic inheritance in different crop species against different pathogens

Disease	Crop	Reference
Hill bunt *(Tilletia foetida)*	Wheat	Luthra and Chandra (1983)
Soil-borne mosaic virus	Wheat	Merkle and Smith (1983), Modawi et al. (1982)
Covered smut *(Ustilago hordei)*	Wheat	S.N. Srivastava and D.P. Srivastava (1977)
Stem rust *(Puccinia graminis)*	Wheat	Sanghi (1974), R. Jain and Gandhi (1978, 1980), Knott and McIntosh (1978), N.D. Williams and Miller (1982), Knott (1984)
Yellow rust *(P. striiformis)*	Wheat	Grama and Gerechter-Amitai (1974)
Powdery mildew *(Erysiphe graminis)*	Wheat	Sebastian et al. (1983)
Leaf blight *(Alternaria triticina)*	Wheat	Sokhi et al. (1973)
Common root rot *(Cochliobolus sativus)*	Wheat	Larson and Atkinson (1981, 1982)
Crown rust *(P. coronata)*	Italian ryegrass	Wilkins (1975)
Stem rust *(P. graminis)*	Barley	Steffensen et al. (1984)
Leaf rust *(P. hordei)*	Barley	Narula et al. (1983)
Blight *(Drechslera sativa)*	Barley	Luthra and Rao (1973)
Blight *(D. maydis)*	Maize	Faluyi and Olorode (1984)
Southern rust *(Puccinia polyspora)*	Maize	Futrell et al. (1975)
Maize dwarf mosaic virus	Maize	Roane et al. (1983)
Grassy stunt virus	Rice	Khush and Ling (1974), Nuque et al. (1982)
Bacterial leaf blight *(Xanthomonas campestris* pv. *oryzae)*	Rice	D.P. Singh and Nanda (1975), Librojo et al. (1976), Petpisit et al. (1977), Yoshimura et al. (1983), R.J. Singh et al. (1983)

Table 4.1 (continued)

Disease	Crop	Reference
Crown rust *(P. coronata)*	Oat	Wong et al. (1983)
Rust *(Puccinia substriata* var. *indica)*	Pearl Millet	Hanna et al. (1985)
Soybean mosaic virus	Soybean	Provvidenti et al. (1982), Buzzell and Tu (1984)
Cowpea aphid-borne mosaic virus	Soybean	Provvidenti et al. (1983)
Blackeye cowpea mosaic virus	Soybean	Provvidenti et al. (1983)
Phytophthora rot *(P. megasperma)*	Soybean	Kilen and Keeling (1981)
Rust *(Phakopsora pachyrhizi)*	Soybean	Bromfield and Hartwig (1980), Hartwig and Bromfield (1983)
Stem rot *(Didymella lycopersici)*	Tomato	Boukema (1982)
Foot and root rot *(Phytophthora nicotianae)*	Tomato	Boukema (1983)
Blight *(Ascochyta rabiei)*	Chickpea	Vir et al. (1974), K.B. Singh and Reddy (1983)
Wilt *(Fusarium oxysporum)*	Chickpea	Sindhu et al. (1983)
Blight *(Ascochyta pinodella)*	Peas	Rastogi and Saini (1984)
Pea seed borne mosaic virus	Peas	Hagedorn and Gritton (1973)
Powdery mildew *(Erysiphe polygoni)*	Peas	J.K. Saxena et al. (1975), Ram et al. (1981)
Pea seed borne mosaic virus	Lentils	Haddad et al. (1978)
Angular leaf spot *(Isariopis griseola)*	French bean	A.K. Singh and Saini (1980)
Bean yellow mosaic virus	Cowpea	Reeder et al. (1972)
Cowpea chlorotic mottle virus	Cowpea	K.M. Rogers et al. (1973)
Cowpea mosaic virus	Cowpea	Raj and Patel (1978)
Cucumber mosaic virus	Cowpea	Khalf-Allah et al. (1973)
Cercospora leaf spot *(C. cruenta)*	Cowpea	Fery et al. (1976)
Bacterial blight *(Xanthomonas vignicola)*	Cowpea	D. Singh and Patel (1977)
Bacterial pustule *(Xanthomonas* sp.)	Cowpea	Patel (1978)
Black eye cowpea mosaic	Cowpea	Taiwo et al. (1981)
Mungbean yellow mosaic virus (tolerance)	Mungbean	Thakur et al. (1977)
Cercospora leaf spot *(C. canescens)*	Mungbean	Thakur et al. (1977)
Bacterial leaf spot *(Xanthomonas phaseoli)*	Mungbean	Thakur et al. (1977)
Cucumber mosaic virus	Mungbean	Sittiyos et al. (1979)
Yellow vein mosaic virus	Okra	Jambhale and Nerkar (1981)
Wilt *(Fusarium oxysporum)*	Alfalfa	Hijano et al. (1983)
Alternaria blight *(A. brassicae)* and *A. brassicola)*	*Brassica juncea*	Tripathi et al. (1978)
White rust *(Albugo candida)*	*B. napus*	Fan et al. (1983)
Clubroot *(Plasmodiophora brassicae)*	*B. campestris*	R.V. James and Williams (1980)
	B. oleracea spp. *capitata*	Chiang and Crete (1983)
Clubroot *(P. brassicae)*	*B. napus*	M.S. Chiang and Crete (1976)
Wilt *(Fusarium oxysporum)*	Sesame	Selim et al. (1976b)
Root rot *(Rhizoctonia solani)*	Sesame	Serry et al. (1976c)
Root and stem rot *(Sclerotium bataticola)*	Sesame	Selim et al. (1976a)

Table 4.1 (continued)

Disease	Crop	Reference
Wilt *(Verticillium alboatrum)*	Sunflower	Fick and Zimmer (1974)
Root rot *(Phytophthora drechsleri)*	Safflower	C.A. Thomas (1976)
Lettuce mosaic virus	Safflower	C.A. Thomas (1981)
Powdery mildew	Linseed	Badwal (1975)
Angular leaf spot *(Pseudomonas lachrymans)*	Cucumber	Dessert et al. (1982)
Downy mildew *(Bremia lactucae)*	Lettuce	Yuen and Lorbeer (1983)
Fusarium yellows *(Fusarium oxysporum)*	Celery	T.J. Orton et al. (1984)

Table 4.2. The oligogenic inheritance in different crop species against different pathogens

Disease	Crop	Reference
Stem rust *(P. graminis)*	Wheat	N.D. Williams et al. (1979), Gough et al. (1974), R. Jain and Gandhi (1983)
Brown rust *(P. recondita)*	Wheat	Hagges and Dyck (1973), Statler (1973), Rasid et al. (1976), Dyck and Samborski (1982)
Leaf blight *(Alternaria triticina)*	Wheat	Sokhi et al. (1973), Kulshrestha and Rao (1976)
Blight *(Drechslera sativa)*	Wheat	O.P. Srivastava (1982)
Leaf blotch *(Septoria tritici)*	Wheat	Rosielle and Brown (1979)
Crown rust *(Puccinia coronata)*	Oat	Wong et al. (1983)
Powdery mildew *(Erysiphae graminis)*	Barley	Moseman and Jorgenson (1973)
Leaf rust *(P. hordei)*	Barley	Narula et al. (1983)
Stripe rust *(P. striiformis)*	Barley	Dayani and Bakshi (1978a)
Loose smut *(Ustilago nuda)*	Barley	K.B.L. Jain and Upadhyay (1974)
Eye spot *(Kabatiella zeae)*	Maize	Reifschneider and Arny (1983)
Northern leaf blight *(Drechslera turcica)*	Maize	A.L. Hooker and Tsung (1980)
Common rust *(P. sorghi)*	Maize	S.K. Kim and Brewbaker (1977)
Sorghum rust *(P. purpurea)*	Sorghum	Rana et al. (1976)
Rice tungro virus	Rice	Seetharaman et al. (1976)
Yellow mosaic virus	Soybean	B.B. Singh and Malick (1978)
Stem canker *(Diaporthe phaseolorum)*	Soybean	Kilen et al. (1985)
Mungbean yellow mosaic virus	Mungbean	Shukla et al. (1978)
Mungbean yellow mosaic virus	Urd bean	D.P. Singh (1980), Dwivedi and D.P. Singh (1985)
Bacterial leaf spot *(Xanthomonas phaseoli)*	Dolichos	Sulladmath et al. (1977)
Sterility mosaic virus	Pigeon pea	H.C. Sharma et al. (1984)
Wilt *(F. oxysporum)*	Alfalfa	Hijano et al. (1983)
Phytophthora root rot *(P. megasperma)*	Alfalfa	Irwin et al. (1981)
Downy mildew *(Bremia lactucae)*	Lettuce	A.G. Johnson et al. (1977)
Wilt *(F. oxysporum)*	Chickpea	Upadhyaya et al. (1983a,b)

Table 4.2 (continued)

Disease	Crop	Reference
Early blight	Tomato	Barkshale and Stoner (1977)
Club root *(P. brassicae)*	Rapeseed	R.V. James and Williams (1980)
Wilt *(F. oxysporum)*	Sesame	Selim et al. (1976b)
Root and stem rot *(Sclerotium bataticola)*	Sesame	Selim et al. (1976a)
Gray mold *(Botrytis cinerea)*	Kenaf	Campbell (1984)
Anthracnose *(Elsinoe ampelina)*	Grapevines	Mortensen (1981)

Table 4.3. The polygenic inheritance in different crop species against different pathogens

Disease	Crop	Reference
Stem rust *(P. graminis)*	Wheat	Knott (1979, 1982)
Brown rust *(P. recondita)*	Wheat	Kumar et al. (1982)
Glume blotch (*Leptosphaeria nodorum* (imperfect stage *S. nodorum*))	Wheat	L.R. Nelson and Gates (1982)
Common bunt *(Tilletia carries)*	Wheat	Metzger et al. (1979)
Barley yellow dwarf	Italian ryegrass	Wilkins and Catherall (1977)
Crown rust *(P. coronata)*	Italian ryegrass	Wilkins (1975)
Rynchosporium orthosporum	Italian ryegrass	Wilkins (1975)
Downy mildew *(Peronosclerospora philippinensis)*	Maize	Kaneko and Aday (1980)
Brown stripe downy mildew *(Sclerophthora rayssiae)*	Maize	I.S. Singh and Asnani (1975)
Southern corn leaf blight *(Drechslera maydis)*	Maize	Lim (1975a,b)
Stalk rot *(Cephalosporium acremonium)*	Maize	Khan and Paliwal (1979)
Stalk rot *(Colletotrichum graminicola)*	Maize	Carson and A.L. Hooker (1981)
Smut *(Ustilago maydis)*	Maize	A.L. Hooker (1978)
Diploidia stalk rot *(D. maydis)*	Maize	A.L. Hooker (1978)
Ear rot *(Fusarium* species)	Maize	Odiemah and Manninger (1982)
Charcoal rot *(Macrophomina phaseolina)*	Sorghum	Rana et al. (1982), Indira et al. (1983)
Grain mould *(Curvularia lunata)*	Sorghum	Dabholkar and Baghel (1980)
Fusarium grain mould	Sorghum	Narayana and Prasad (1983)
Downy mildew *(Sclerospora graminicola)*	Pearl millet	Basavarajn et al. (1981)
Bacterial blight (*Xanthomonas campestris* pv. *malvacearum*)	Cotton	Innes et al. (1974)
Chocolate spot *(Botrytis fabae)*	Broadbean	J.E.M. Elliot and Whittington (1979)
White mould *(Sclerotinia sclerotiorum)*	Snapbean	Fuller et al. (1984)
Rhizoctonia root rot	Snapbean	M.H. Dickson and Boettger (1977)
Pythiium seed rot and damping off *(P. ultimatum)*	Snapbean	York et al. (1977)
P. myriotylum	Snapbean	Reeleder and Hagedorn (1981)

Table 4.3 (continued)

Disease	Crop	Reference
Haloblight *(Pseudomonas phaseolicola)*	Snapbean	Coyne and Schuster (1974)
Black rot *(Xanthomonas campestris* pv. *campestris)*	Cauliflower	B.R. Sharma et al. (1972)
Fusarium yellows *(Fusarium oxysporum)*	Celery	T.J. Orton et al. (1984)
Fusarium oxysporum	Tulip	Vaneijk and Eikelboom (1983)

polygenic (Table 4.3). In the case of monogenic type of resistance, the reports suggest that the dominance of resistance and susceptibility are equal. As compared to the literature published before 1971, more reports are now available on the oligogenic and polygenic types of resistance. More reports are available on the allelic relationships of res genes, particularly in crops like rice and wheat, although relatively few findings deal with the linkage relationships of res genes.

The inheritance of resistance could be studied by the use of genetically controlled cultures and the homozygous resistant and susceptible host lines. The parents, F_1's, F_2's, back-crosses and/or F_3 progenies, are inoculated with single-spore cell culture. The scoring of individual plants is recorded. The plants/progenies are classified based on the reactions of parents and F_1's.

4.2 Inheritance of Induced Resistance

The inheritance of induced resistance may be dominant, as in the case of stem rust resistance in wheat and oats, or recessive, as in the case of *Victoria* blight resistance in oats (Konzak 1956). Luke et al. (1966) noted that sensitivity of oats mutant to *Drechslera (Helminthosporium)* blight was directly related to the necrotic reaction to crown rust, confirming that the reaction to both diseases is controlled by the same gene. The induced mutants, SR 1, SR 7 and SR 10 of barley show almost immune reaction to powdery mildew, *Erysiphe graminis* f. sp. *hordei* (Wiberg 1973). Genetic analysis has shown that each of these has a single, recessive factor conditioning resistance, and that these factors are heteroalleles. A mutant rice strain MR 515–17 was selected after irradiation of variety Plakweng which is susceptible to blast *(Pyricularia oryzae)* disease. The resistance of this mutant strain to a new race IA–65 of *P. oryzae* appeared to be controlled by a dominant gene (Jenng 1975).

Induced resistance is also reported to be due to minor genes in rice to blast disease through radiation treatment (Lin and Lin 1960). Soybean seeds irradiated wih γ-ray and EMS-treated showed induced mutants for rust *(Phakopsora pachyrhizi)* was due to the minor genes (Tsai et al. 1974).

4.3 Genetics of Pathogenicity

During the period following Biffen's work, the knowledge of the inheritance of resistance to diseases was being expanded, parallel advances were being made towards the understanding of variation in fungi. Early studies on the inheritance of virulence were carried out in the early 1930's with the fungus *Ustilago* (Nicolaisen 1934). This author showed that the capacity to incite a host and the expression of either a susceptible or a resistant disease reaction was under Mendelian control. Van der Plank (1968) noted that virulence may be inherited as oligogenic and aggressiveness is polygenic. Virulence involves gene diversity probably largely through mutations. Aggressiveness may well involve enzyme dose (as distinct from enzyme diversity) and the switching on and off of enzyme action. Vanderplank (1975) observed that virulence and avirulence in the pathogen are the counterparts of vertical susceptibility and resistance in the host. Aggressiveness and unaggressiveness in the pathogen are the counterparts of horizontal susceptibility and resistance in the host.

Person and Sidhu (1971) reviewed the literature on the genetics of pathogenicity and made the following generalizations:

– The virulence/avirulence was usually under Mendelian control.
– The genes which induced susceptible reaction were usually inherited as recessives.
– Linkage of genes for virulence was reported only occasionally and there was no report on allelism.

Only few selected findings are described here to suggest that inheritance of virulence could be monogenic or polygenic.

R.R. Nelson and Kline (1969) studied the pathogenicity of 291 ascospore isolates obtained from different crosses between isolates of *Cochliobolus heterostrophus (Drechslera maydis)* to nine differential gramineous species in 37 differential segregations. A minimum of 13 different genes for pathogenicity was identified. Segregation ratios and comparisons of responses of paired differential species indicated that pathogenicity to four host species is controlled by five different genes, and the pathogenicity to four species is conditioned by five different sets of two genes each. All pathogenic capacities are inherited independently. The studies illustrate the genetic complexity for pathogenicity in *C. heterostrophus*.

Sidhu and Person (1971) reported that virulence of *Ustilago hordei* isolates is recessive and governed by a single gene. Monogenic inheritance of virulence was also observed by Lim et al. (1974) in the case of *D. turcica* isolates.

Drechslera sativa isolated from wheat and barley (non-pathogenic to each other source) were crossed on Sachs' medium. The monoscosporic progenies segregated in 1:1:1:1 ratio, i.e. 25% progenies were pathogenic to wheat, 25% to barley, 25% to both and 25% to none (R.J. Singh et al. 1983). The two isolates of *D. maydis* varying in their pathogenicity (linked with mycelial colour) were crossed. They segregated in the ration of 1:2:1 for pathogenic (dark grey colour), moderate (intermediate colour) and non-pathogenic (white colour), respectively. Maize and teosinte isolates, non-pathogenic to each other, showed the segregation ratio of 1:1:1:1 for pathogenic to maize, teosinte, both maize and teosinte, and to maize, respectively (R.J. Singh et al. 1983).

Crosses between isolates of *Gaeumannomyces graminis* var. *tritici* differing in pathogenicity were made using a weakly pathogenic auxotrophic UV-irradiated mutant as
one of the parental isolates in each cross. Pathogenicity is under multiple gene control
and most genotypic variation is additive (Blanch 1980).

A hybrid between two biotypes of *Ustilago nuda* produced segregating progeny
that were used to identify two genes for virulence on five cultivars of barley. Gene
Unv_1 was responsible for virulence on Warrior, Compana, and Valkie, while gene Unv_2
was responsible for virulence on Keystone and Bonanza. The two genes were recessive
and inherited independently from each other and from the loci for mating type and
spore colour (P.L. Thomas 1982).

Inheritance of virulence can be studied by crossing the isolates/races with differing
pathogenicity. The parents, F_1, F_2 etc. are inoculated on the tester host genotype.
The method of selfing F_1 rust cultures was reported by Miah (1968). He observed
that the pooled nectar method of selfing F_1 culture introduces sampling errors because
of limited number (a) of pycnia selected, (b) of aecia sampled, and (c) of cultures
derived per aecium. The reciprocal method of selfing is more reliable.

4.4 Gene-for-Gene Concept

The gene-for-gene hypothesis was proposed by Flor (1942, 1955) as the simplest ex-
planation of the results of studies on the inheritance of pathogenicity in the flax rust
fungus, *Melampsora lini*. On the varieties of flax *(Linum usitatissimum)* that had one
gene for resistance to the parent race, F_2 cultures of the fungus segregated into mono-
factorial ratios. On varieties which had 2, 3 or 4 genes for resistance to the parent race,
the F_2 cultures segregated into bi-, tri-, or tetrafactorial ratio (Flor 1942, 1947). This
suggested that for each gene that conditions resistant reaction in the host there is a
corresponding gene in the parasite that conditions pathogenicity. Each gene in either
member of a host-parasite system may be identified only by its counterpart in the
other member of the system.

The correspondance between the number of genes for resistance in the host and the
number of recessive genes for pathogenicity in the rust is presented in Table 4.4. It

Table 4.4. Reaction of flax varieties to F_2 segregates of a hybrid between race 22 and 24 of *Melampsora lini*. (Flor 1959)

Variety of flax	No. of resistance genes in the host	F_2 ratio of segregates of rust		No. of virulence genes in race 22 and 24
		Pathogenic	Non-pathogenic	
Ottawa 770B	1	1	3	1
Newland	1	1	3	1
Bombay	1	1	3	1
Akmolinsk	1	1	3	1
Italia Roma	2	1	15	2
Bolley Golden	2	1	15	2
Morye	3	1	63	3

may be emphasized, however, that the total number of genes concerned with infection in both the parasite and the host is not necessarily the number indicated by these tests. For instance, a variety of flax may have two dominant genes for resistance, say LLNN, and may be susceptible to one race of the rust and resistant to another. These races could have the genetic constitution $a_L a_L a_N a_N$ and $A_L A_L a_N a_N$. The hybrid between these two races would segregate to give a ratio of 3(non-pathogenic): 1(pathogenic) isolates in the F_2 generation when tested on the tester with LLNN genotype. This shows that a hybrid between two races yielding a monogenic segregation on a digenic host variety.

The results obtained by Flor allowed the conclusion that for every gene for resistance in the host, a given pathogenic race must possess complementary alleles to enable it to overcome the resistance. This complementary relationship of host and pathogen is called the gene-for-gene hypothesis for disease resistance.

A further example of the operation of the gene-for-gene hypothesis system in host and parasite is given in Tables 4.5 and 4.6. The two varieties Ottawa 770 B and Bombay, each having one dominant gene for resistance, and their hybrid segregated in the F_2 generation to give a digenic ratio when tested with two rust races 22 and 24. Similarly,

Table 4.5. Digenic inheritance of resistance in flax varieties. (Flor 1956a)

Rust race	Genotypes of parental flax varieties		F_2 genotypes			
	Ottawa 770B (LLnn)	Bombay (llNN)	LN	Ln	lN	ln
22	S	R	R	S	R	S
24	R	S	R	R	S	S
		Nos. obeserved	110	32	43	9
		Expected on the basis of 9:3:3:1 ratio	108	36	36	12

Table 4.6. Digenic inheritance of pathogenicity in races of *Melampsora lini*. (Flor 1956a)

Flax variety	Genotypes of parental races		Genotypes of rust segregates in F_2			
	Race 22 ($a_L a_L A_N A_N$)	Race 24 ($A_L A_L a_N a_N$	$A_L A_N$	$a_L A_N$	$A_L a_N$	$a_L a_N$
Ottawa 770B (LLnn)	S	R	R	S	R	S
Bombay (llNN)	R	S	R	R	S	S
	Nos. observed		78	27	23	5
	Expected on the basis of 9:3:3:1 ratio		75	25	25	8

R = resistant; S = susceptible

the F_2 segregates of the hybrid between races 22 and 24 showed a digenic ratio when tested on the varieties Ottawa 770 B and Bombay.

Resistance genes occur as multiple alleles in five loci (Flor 1971). The symbols L, M, K, N and P designate dominant genes in five loci. Three of the loci K, L and M appear to be independently inherited, while N and P are located on the same chromosome at a distance of about 26 cross-over units. Of the 26 resistance genes that have been identified to five independent loci are K (2), L (11), M (6), N (3), and P (4). The figures in parenthesis are the number of alleles. K.M.S. Saxena and Hooker (1968) reported 15 alleles at locus Rp1 in maize which determine resistance to common rust *(Puccinia sorghi)*. These alleles are supposed to have originated through tandem duplications.

Genes conferring host resistance to an obligate parasite grouped together in complex loci provide opportunities to study their structure. A modified cis-trans test was used to interpret the position effects of codominant genes mutually recombined within each of the two complex loci (M and L) with the use of specifically developed method of analysis among F_2 segregates (Shepherd and Mayo 1972).

Flor has suggested genetic symbols for the rust races which serve to indicate directly their relationship to the resistance genes in the host. Pathogenicity in rust is symbolized by the letter A or a – A denoting avirulence (non-pathogenic dominant allele) and a virulence (pathogenic recessive allele). Each locus in rust is further identified by subscripts, indicating the locus and allele of the host variety to which it is related. For instance, the allele in rust causing pathogenicity on a flax genotype of the construction N^1 is designated a_N1 and the non-pathogenic allele at the same locus as A_N1. The system is illustrated in Table 4.7.

Only the rust races possessing recessive alleles for pathogenicity in respect of all the dominant resistance genes present in the host genotype are capable of producing infection.

The gene-for-gene hypothesis was based on correlated genetic studies of both host and parasite (Flor 1942, 1955). Acceptance of the hypothesis enables one to construct hypothetical genotypes of host and pathogen in the absence of direct genetic studies by determining the reaction of a range of host varieties to a range of pathogen races (Person 1959). Statler et al. (1981) used hybridization of *M. lini* to identify rust resistance in flax. Selected races of *M. lini* were crossed and selfed to produce virulent patterns to identify combinations of resistance alleles L^{11}, M^3, M^6 and P^3 relatively simple. Virulence and resistance may act on the same biochemical mechanisms. *Erwinia* virulence on potato depends on the lysis of cell walls of the host, resistance may depend on the lysis of cell walls of the parasite (Nitzsche 1983).

Table 4.7. Relationship between the genotypes of flax and of flax rust. (Flor 1956a)

Flax variety	Genotype of flax variety	Genotype of rust races	Host reaction
Polk	N^1N^1pp	$A_N1\ A_N1\ a_pa_p$	Resistant
Polk	N^1N^1pp	$a_N1\ a_N1\ A_pA_p$	Susceptible
Koto	nn PP	$a_N1\ a_N1\ A_pA_p$	Resistant
Koto	nn PP	$A_N1\ A_N1\ a_pa_p$	Susceptible
Redwood	N^1N^1PP	$A_N1\ A_N1\ a_pa_p$	Resistant
Redwood	N^1N^1PP	$a_N1\ a_N1\ a_pa_p$	Susceptible
Winona	nnpp	A_N1 or a_N1, A_p or a_p	Susceptible

Table 4.8. Host-parasite systems for which gene-for-gene relationships have been demonstrated or suggested

	System	Reference
Rust	*Zea–Puccinia sorghi*	Flangas and Dickson (1975), A.L. Hooker and Russell (1962)
	Linum–Melampsora lini	Flor (1942, 1955, 1956a)
	Triticum–Puccinia graminis	I.A. Watson and Luig (1958), Loegering and Powers (1962), Green (1966) Kao and Knott (1969)
	Triticum–P. striiformis	Zadoks (1961)
	Triticum–P. recondita	Samborski and Dyck (1968)
	Avena–P. graminis	Martens et al. (1970)
	Coffea-Hemileia vastatrix	Noronha-Wagner and Bettencourt (1967)
	Helianthus–P. helianthi	Sackston (1962)
Smuts	*Avena–Ustilago avenae*	Holtan and Halisky (1960)
	Triticum–U. tritici	Oort (1963)
	Hordeum–U. hordei	Sidhu and Person (1972)
Bunts	*Triticum–Tilletia caries*	Metzger and Trione (1962)
	T. contraversa	Holton et al. (1968)
Mildews	*Hordeum-Erysiphe graminis*	Moseman (1957, 1959)
	Triticum-E. graminis	Powers and Sando (1957), Slesinski and Ellingboe (1970)
	Medicago-Peronospora trifoliorum	Skinner and Stuteville (1985)
Apple scab	*Malus–Venturia inaequalis*	Boone and Keitt (1957), P.K. Day (1960), Bagga and Boone (1968)
Late blight	*Solanum–Phytophthora infestans*	Toxopeus (1956)
Blight	*Zea–Drechslera turcica*	Lim et al. (1974)
Leaf mould	*Lycopersicon–Cladosporium fulvum*	P.R. Day (1956)
Potato wart	*Solanum–Synchytrium endobioticum*	Howard (1968)
Bacteria	*Gossypium–Xanthomonas campestris* pv. *malvacearum*	Brinkerhoff (1970)
	Leguminosae–Rhizobium (Symbiosis)	Nutman (1969)
Viruses	*Lycopersicon*–Tobacco mosaic virus	Pelham (1966)
	Lycopersicon–Spotted wilt virus	P.K. Day (1960)
	Solanum–Potato virus	Howard (1968)
	Phaseolus–Bean common mosaic virus	Drijfhout (1978)

The usefulness of Flor's hypothesis in interpreting the results of genetic studies of parasitism is illustrated in Table 4.8, which lists the systems in which the operation of the gene-for-gene relationships has been either demonstrated or suggested.

4.5 A Second Gene-for-Gene Hypothesis

Flor's gene-for-gene hypothesis is purely a hypothesis of identities. The resistance gene in the host and the corresponding virulence gene can be identified by this hypothesis. But it does not tell us anything about the gene quality. A second gene-for-gene hypothesis which assumes Flor's hypothesis tell us about the quality of resistance genes.

The quality of resistance gene in the host determines the fitness of the matching virulence gene in the parasite to survive, when this virulence gene is unnecessary. "Unnecessary" gene means that the matching resistance gene in the host is not present. Reciprocally, the fitness of the virulence gene in the parasite to survive when it is unnecessary determines the quality of matching resistant gene in the host, as judged by the protection it can give to the host. For instance, there are ten or more genes for resistance to late blight of potato, $R_1, R_2, ... R_{10}$. These genes can be identified. Of these, the first four R_1, R_2, R_3 and R_4 have been well studied. Are these genes of equal importance and do they have same effect? The answer is definitely no. From the reports published in literature, we find that the R_4 gene has not been successfully used on its own by breeders, although it has been occasionally used in combination with other resistant genes. The R_1 gene has often been used on its own and it has protected potato varieties against blight successfully. The difference between these two R genes is explained by the fact that virulence on R_4 preexisted abundantly in populations of *P. infestans*, whereas virulence on R_1 did not (Vanderplank 1975b).

Hogen-Esch and Zingstra (1957) compiled a list of potato varieties, giving information among other things about R genes. None of the varieties in the list has the R_4 gene alone; however, it has been used in five varieties in combination with other R genes. In contrast, the gene R_1 has been used on its own in 54 varieties and in combination with other R genes (mostly R_3) in 19 varieties. This indicated that gene R_4 needed the help of other genes more than R_1 did.

Frandsen (1956) collected 34 isolates of late blight *(P. infestans)* from farmers' fields in north western Germany in the early 1950's. The potato varieties cultivated in that region had no R genes, yet all isolates were virulent on R_4 gene, suggesting that virulence preexisted. Potato fields in Canada had no R_4; this gene has not been used commercially there. Yet out of 68 isolates collected from farmers' fields, 39 were virulent on R_4. This also suggested that virulence preexisted abundantly enough to make the gene R_4 practically worthless to potato breeders (K.M. Graham 1955). These studies indicated that the great fitness of virulence on R_4 to survive even when it is unnecessary shows that the quality of R_4, judged by the protection it can provide, is poor.

The ratio for virulence between R_1 and R_4 genes differs significantly. Therefore, it proves that there is a difference in the quality of resistance genes R_1 and R_4 as shown by the greater use of R_1 by the potato breeders and the greater fitness of virulence on R_4 to survive when it is unnecessary.

4.6 The Gene-for-Gene and the Protein-for-Protein Hypothesis

In diseases in which host and pathogen are involved gene-for-gene, susceptibility involves the copolymerization of protein from the host with protein from the pathogen (Vanderplank 1978).

In gene-for-gene diseases host and pathogen recognize each other by their proteins. This hypothesis was stated by Vanderplank (1976). He stated that the protein-for-protein hypothesis is applied to diseases to which the gene-for-gene hypothesis has been applied. The two hypotheses differ some-what in emphasis. The gene-for-gene hypothesis centres around genes for resistance in the host plants. The protein-for-protein hypothesis is concerned primarily, but not exclusively, with susceptible host plants, i.e. with compatible host-pathogen combinations.

The known classes of substances capable of storing information of variation are nucleic acids, proteins and complex carbohydrates including the glycoproteins (Vanderplank 1978). Proteins are placed in the centre of the sequence of synthesis, and receive most of the information on variations from the nucleic acids and are vastly more versatile and reactive than the nucleic acids in making use of the information. They filter out most of the information on qualitative variation, and make their products relatively poor storehouses of genetic variation. The polymerization serves to bring the protein-stored information from the host and the pathogen together without loss through catalysis. Host-protein and pathogen-protein must make physical contact with each others, without their products being intermediaries, i.e. without the loss of variation that catalysis involves. The copolymer is derived partly from the host and partly from the pathogen, which is what the protein-for-protein hypothesis is about (Vanderplank 1978).

Vanderplank (1978) hypothesized that in susceptibility the pathogen excretes a protein into the host cell which copolymerizes with a complementary host protein. This copolymerization interferes with the autoregulation of the host gene that codes for the protein, and by so doing turns the gene on to produce more protein. This host-protein in fungal disease serves primarily as food for the pathogen; it is the essence of parasitism that the host feeds the parasite. The copolymerization has a second function, i.e. it binds the protein entering the host from the pathogen, and takes it out of circulation as a catalyst.

In resistance, the protein specified by the gene for avirulence in the pathogen and excreted into the host does not polymerize. It stays actively catalytic in the host cell as a foreign body, introducing catalytic processes not normal to the healthy host cell. It is the "elicitor" or "toxin" that catalyzes the start of the reactions that back-lash and inactivate the pathogen, directly or indirectly, as by starvation.

The same protein from the pathogen, isozyme variations apart, can, if copolymerized, make for susceptibility, and turn on a supply of food for the pathogen, or, if unpolymerized, act to promote resistance reactions.

4.7 Biochemical Basis of the Gene-for-Gene Hypothesis

It has been speculated by plant pathologist that the plant host parasite might have an immunological basis analogous to that between animals and their parasites. A more interesting experiment was conducted by Doubly et al. (1960), who followed the gene-for-gene hypothesis with a serological study of host-pathogen relations in the flax *M. lini* system. They used four varieties of flax with varying resistance and four isolates of *M. lini* with varying virulence. As antigens they used globular protein fractions from the flax varieties and from urediospores of the rust isolates. They prepared antisera in rabbits. The host-pathogen interaction was susceptible, when titres of rust antiserum against flax antigens were relatively high (1:160 or 1:320) and resistant when titres were low (1:20 or 1:40).

Schnathorst and De Vay (1963) attempted similar study in the *Gossypium hirsu-tum-X. campestris* pv. *malvacearum* system. Three cotton varieties with different resistance reactions were used. When susceptibility occurred, strongest antigen-anti-serum reaction could be found.

A biochemical mechanism for the gene-for-gene resistance of tomato to *Cladosporium fulvum* was carried out by Van Dijkman and Sijpesteijn (1971). They observed that a gene-for-gene relation existed between tomato and *C. fulvum*. It was reported to be based on the interaction of specific fungal excretion products with the receptors in the host which may be located in the cell membrane. The fungal compounds were noted to be controlled by four avirulent genes (A_1, A_2, A_3 and A_4) and that of the receptors by four resistance genes (Cf_1, Cf_2, Cf_3 and Cf_4).

RNA is directly involved in the resistant reaction of wheat stem rust and the active RNA shows specificity comparable to that of the host/parasite system (Rohringer et al. 1974). The antibody/antigen relationship may exist between host cell contents and nematode saliva (F.G.W. Jones 1974).

De Vay et al. (1972) have generalized that tolerance of parasites by the host increases with increasing antigen similarity, whereas resistance of the host is characterized by an increasing disparity. Tolerance is used to mean the submission of the host to the parasite, thus tolerating its attack.

Successful parasitism depended on a substantial quantity of antigens being shared by host and pathogen. In other words, greater disparity between antigens of host and parasite led to vertical resistance (Vanderplank 1975b). Antigens common to host and parasite would imply genetic material common to host and parasite and antigenic disparity would imply genetic disparity.

4.8 How Does the Gene-for-Gene Relationship Come into the Picture?

Host and parasite (obligate) must have evolved together. Mutations that confer an advantage to the host must be disadvantageous to the parasite and vice versa, if the relationship between host and parasite is antagonistic. The two mutational events, i.e. mutation to resistance in the host and mutation to virulence in the parasite have the same value in the evolution of host and parasite system. As a result of such genetic interaction between host and parasite, both participants have developed a far-reaching biological and physiological specialization. Pathogens have tended to be selective among their hosts, and hosts have developed resistance against their special parasites (Gentry 1969).

The genes for resistance have value only in a susceptible population. Unless the resistance gene is in the population, the new mutation to virulence in the parasite will have no selective advantage. Thus the general rule emerges that of the two main evolutionary events that occur in the host-parasite system, the increase in gene frequency of host genes for resistance occurs first and is followed by increase in the frequency of genes for virulence that overcome their effect. The mutation for resistance can occur at one or more than one locus, but having occurred at one locus, the mutant gene for resistance has two effects: (1) Selection pressure will be removed at other loci in the host that are capable of supplying resistance genes. (2) Selection pressure will be improved at all loci in the parasite capable of supplying an effective gene for virulence.

Considering the parasite, if an effective mutation for virulence occurs at particular loci, its action will be related to the specific gene in the host where the effect is now overcome. A gene-for-gene relationship has thus come into the picture. Thus there is a dynamic system of gene-for-gene relationship in host-parasite systems.

4.9 Genetic Limitations to Models of Specific Interactions of the Host and its Parasite

Person and Mayo (1974) noted that specific interaction between a resistance gene (R gene) of the host and an avirulence gene (A gene) of the parasite is the basis of the gene-for-gene relationship. R and A genes are conditional genes, capable of associating with either of the two phenotypes, depending on the presence or absence of the other. Although avirulence usually is a dominant character, micro-evolutionary history suggests that in its primary function an A gene associates not with avirulence but with virulence, also that R:A interaction generates a "stop signal" which prevents disease development, and finally that the primary function of a recessive a-gene, when it replaces a dominant A-gene, is one of negating the stop signals. Where several gene-for-gene relationships operate simultaneously, a single R:A interaction is sufficient to prevent disease development. R and A genes thus occupied in R:A interaction are epistatic over those which are not. Because of the characteristics of the gene-for-gene relationship, definite limitations are imposed on genetic study. Because of these limitations in any genetic study these may be contributing genes in both the host and the parasite which are not detected.

4.10 Gene-for-Gene Basis in HR

The inheritance of mature plant resistance to rust *(Puccinia sorghi)* in corn appears to be due to numerous genes. Based on calculated heritability, this type of resistance is highly heritable (A.L. Hooker 1967).

The partial resistance of barley *(Hordeum vulgare)* to leaf rust *(Puccinia hordei)* shows all the characteristics of the so-called HR (Parlevliet 1977). Partial resistance in the field, the latent period and the infection frequency in young flag leaves were measured in three fairly resistant cultivars with five different isolates. The cultivars and isolates differed significantly for all three variables. The cultivar-isolate interaction component was highly significant for the latent period; the Julia-isolate 18 combinations showed a differential interaction variance. In the field, the partial resistance of the Julia-isolate 18 combination also appeared to interact differentially. The cultivar effect on the latent period was shown earlier to be governed by polygenes. This indicates that differential interactions occur in a polygenic systems, and suggests that polygenic systems in the host could operate on a gene-for-gene basis with polygenic systems in the pathogen.

Parlevliet and Zadoks (1977), accepting Vanderplank's definition of HR and VR, did not agree with the latter's ideas about the genetic basis of HR. Through simulation studies, they showed that polygenic resistances operating on a gene-for-gene basis with polygenes in the pathogen could be race non-specific or horizontal in nature. They assumed that most true resistance genes operate on a gene-for-gene basis with genes in the pathogen. Polygenic resistances of a horizontal nature are supposed to be stable because of many genes and not because other genes are involved.

Parlevliet (1978) reported that polygenes indeed can behave as major race-specific or vertical resistance genes do. He observed that partial resistance of barley to leaf rust, (P. hordei), is characterized by a reduced rate of epidemic development in spite of a susceptible infection type. Barley cultivars vary greatly in partial resistance and its components. In a test for interaction between host cultivars and pathogen isolates, most variation was of a horizontal nature.

The inheritance of slow leaf-rusting resistance of the wheat (T. aestivum) cultivar Suwon 85 was studied in crosses with susceptible cultivars Moon and Suwon 92. Plants were grown in the glasshouse and inoculated with P. recondita f.sp. tritici. Two components of slow leaf-rusting resistance, latent period (LP) and pustule size, were measured. The longer latent period of Suwon 85 is controlled by two genes, the small pustule size also to some extent. Thus it should be possible to select for these components of slow leaf-rusting resistance in segregating populations (Kuhn et al. 1980).

Wheat varieties UPB 10, Janak, WH 147, GLR-1 and GLR-22 were rated as slow leaf rusting to P. recondita f.sp. tritici in the field because these varieties allowed a lower infection rate as compared to fast-rusting varieties Lal Bahadur, HI 601 and Kalyansona. These slow-rusting varieties showed a longer incubation period, fewer uredospores/pustules and early conversion of uredia into telia on flag leaves as compared to fast trusting Lal Bahadur and all other varieties tested under glasshouse conditions (R.P. Gupta and A. Singh 1982).

The study conducted by Villareal (1980) on different components of HR in rice to rice blast disease revealed an interaction between HR and epidemiological fitness, indicating that HR could possibly erode to some extent over time.

4.11 Additive Effects, Genetic Background, Gene Dosage, Dominance Reversal, Heterotic Effect and Environmental Effects

The chloretic lesion type of resistance in maize to the northern leaf blight (D. turcica) fungus is conditioned by a single dominant gene, Ht (A.L. Hooker 1963). The resistance is expressed as a restriction of growth and sporulation of the pathogen. A second main form of resistance affecting lesion number is under multiple gene control. Both types of resistance qualify as HR when viewed as rate-reducing. Most of the gene action in the lesion number form of resistance is additive (Hughes and A.L. Hooker 1971). The expression of chloretic-lesion resistance is enhanced by a background containing genes for lesion-number resistance.

The gene dosage effects of the Ht alleles was studied by G.M. Dunn and Namm (1970). They obtained monoploid, diploid, triploid and tetraploid seedlings of maize for Ht and ht alleles and inoculated at the three- or four-leaf stage in a humid chamber at 19° to 21 °C. Disease reaction was evaluated by determining the percentage of leaf area infected. There was no difference in resistance between monoploid (Ht) and diploid (HtHt) or between triploid (HtHtHt) seedlings; however, three or four doses of the Ht allele conferred a higher level of resistance on seedlings than did one or two doses.

Heterozygous diploid (Ht/ht) seedlings were always less resistant than any of the other levels of Ht used in this study.

Diploid, triploid and tetraploid seedlings containing two, three and four doses, respectively, of the ht allele did not differ in their degree of susceptibility. Monoploid (ht) seedlings were much more susceptible than seedlings of the other three dosage level. A.L. Hooker (1973) also reported a dosage effect of the Ht alleles. Greater kernel weights and lower leaf blight scores were associated with the presence of Ht gene (A.L. Hooker and Kim 1973). The dosage effects or possibility of such effects have been reported in potato (Ferris 1955, Toxopeus 1957) and in apple (Aldwinckle et al. 1977).

Athwal and Watson (1954) presented evidence that resistance genes have major or minor effects in different genetic backgrounds. When the major gene B_7, conferring resistance to bacterial blight of cotton, was transferred to two susceptible upland varieties, it was found to be much more effective in one variety than in the other (Innes 1964). A line of wheat K58Mg[10], homozygous for gene Sr 6 for resistance to stem rust, was crossed to four stem rust-susceptible lines. Sr 6 was dominant with both race 56 and race 15 B-1 in two crosses, dominant with race 56, and recessive with both races in the fourth cross. It showed that dominance of Sr 6 depended on the susceptible parent with which K 58 Mg[10] had been crossed. The dominance of Sr 6 was not controlled either by a single independent modifier locus or by the general background of the susceptible parents. Although the possibility of a linked modifier has not been eliminated, it appears that the result may be due to the effects of different alleles for susceptibility at the Sr 6 locus. Degree of resistance conditioned by Sr 6 is determined by a complex interaction between genotype and environment (Knott 1981).

Inheritance studies involving inbred, F_1, F_2 and F_3 plants of resistance inbreds (NN14 and M16) and a susceptible inbred in corn showed that each of the resistant inbreds carries a single dominant gene conferring resistance to culture 901 aba of *P. sorghi* and a single recessive gene conferring resistance to culture 933a. The two genes segregated as a unit. Two hypotheses account for the apparent reversal of dominance: the dosage effect, and two closely linked genes (A.L. Hooker and K.M.S. Saxena 1967).

The inheritance of resistance to the bacterial leaf blight (*X. campestris* pv. *oryzae*) isolate PX061 (from the Philippines) in five rice cultivars was studied by Sidhu and Khush (1978). The pattern of segregation suggested a monogenic recessive factor when the plants were inoculated at the booting stage, but it was monogenic dominant when the plants were inoculated during flowering. The dosage effect of the resistance gene causes this reversal of dominance, as the heterozygous plants were susceptible at booting and resistant during flowering. This gene, designated Xa6, is linked to Xa4, another dominant gene for resistance with a cross-over value of 26%.

The heterotic effect of resistance to *Drechslera maydis* race 0 in maize was estimated from a set of diallel crosses in ten parental lines. Disease evaluations, based upon disease rating on a scale of 0 to 100 and lesion length, were made at mid-silk following artificial inoculation (Lim 1975b). Variations attributed to average heterosis, by line, and specific heterosis, between lines, were highly significant. Resistance was partially dominant. Resistant inbreds exhibited resistant line effects, with inbred Pa 884p exhibiting the most resistant effect for both disease evaluations. Differences existed between average heterosis effects of disease evaluations. Based on disease rating, resistant inbreds contributed less heterotic effects for resistance to single crosses than did susceptible inbreds. Susceptible inbred Mo 19 contributed more heterotic effects for resistance to single crosses than any other parental inbreds. Data on lesion length, however, showed that resistant inbreds Pa 884p and Va 43 contributed greater effects of average heterosis for resistance to single crosses than other inbreds. The heterotic effect for lesion length was smaller in magnitude than that for disease rating.

Disease evaluation generally indicated that progeny from crosses between resistant inbreds are more resistant to race 0 than are those involving intermediately resistant or susceptible inbreds.

The rust reaction of seedlings of 87 varieties of wheat to *Puccinia graminis* f.sp. *tritici* races 17, 38 and 56 were determined at greenhouse temperatures of 70° and 85 °F by Bromfield (1961). With respect to a give race, the varieties were, (1) resistant at both temperatures, (2) susceptible at both temperatures, or (3) resistant at 70° but susceptible at 85 °F. Certain varieties that became susceptible at high temperature (temperature-sensitive) were further tested at 70°, 77° and 85 °F. Twenty varieties inoculated with race 17, 12 inoculated with race 38, and 15 inoculated with race 56 were resistant at 70 °F but susceptible at 77° and 85 °F. The reaction of seven temperature-sensitive varieties to races 11, 15B, 17, 38 and 56 were observed within the critical temperature zone in which the breakdown in rust resistance occurs (70°–77 °F). Individual differences in behaviour were noted among the various host-race combinations, but in general varieties were (1) resistant at 70 °F, (2) mixed in reaction at 72–74 °F, and (3) susceptible at 76 °F.

High temperature appears to favour lesion development of bacterial leaf blight in rice (Horino et al. 1981). The infection of seedlings of the wheat Maris Fundin by *P. recondita* f.sp. *tritici* was investigated at three temperatures and compared with that on Armada, which always gave a compatible reaction (Hyde 1982). Maris Fundin gave an incompatible reaction at 20 °C, a compatible reaction at 9 °C, and an intermediate reaction at 15 °C. At 15 °C Maris Fundin developed fewer uredosori and had a longer latent period than Armada.

4.12 The Vertifolia Effect

Van der Plank (1963) has defined vertifolia effect as the loss of horizontal resistance in the process of breeding for vertical resistance. The implications are that in an epidemic covering a wide area for a long time, even a slight increase of HR can be beneficial and even a small loss of HR can be detrimental.

Kirste (1958) collected data on the progress of blight *(P. infestans)* in plots of potato of 12 late-maturing varieties (six with and six without R genes). The data were collected for blight to progress from rating 1 (a very mild attack) to rating 3 (medium infection). In varieties with R genes, the average date for rating 1 and 3 was August 15 and 30, respectively, i.e. a span of 15 days. In the varieties without R genes, the corresponding dates were August 7 and September 2, i.e. a span of 26 days. The start of the epidemic was delayed by R genes, but once started was much faster when R genes were present. The infection rate, r, was in fact 26/15 times faster in varieties with R genes than in those without. Weather conditions that affected blight were the same in varieties with R genes and without R genes.

Five of the six varieties have the gene R_1. Vertifolia has the R_3 and R_4. It had a high infection rate. The blight rating in Vertifolia cultivar was 1 on August 22, suggesting the fact that combined virulence on R_3 and R_4 was very rare in a population of *P. infestans*, but rating 3 was reached on August 29, i.e. 6 days later. The infection rate, r, was 26/6 times faster in Vertifolia than in varieties without R genes (Vanderplank 1978). The data on similar lines were repeated by Schick et al. (1958a,b), who rated as very resistant the cultivar which took 16 days for blight to progress from rating 1 to 3 when R genes were present, but 32 days when R genes were absent. With moderately resistant cultivars the corresponding figures were 12 and 25 days. They assessed that resistance was both vertical (delays the onset of epidemic) and horizontal (slows down epidemic). If one separates out the HR and measures it as being inversely proportional to the time taken for blight to progress from rating 1 to 3, the infection rate, r, was about twice as fast in varieties with R genes as in those without (Vanderplank 1978).

4.13 Host-Specific Phytotoxins and Pathogenesis

Phytotoxins representing a diverse class of chemical compound have been isolated from plant pathogenic bacteria and fungi (Strobel 1974). A host-specific plant toxin is defined as a metabolic product of a pathogenic micro-organism which is toxic only to the host of the pathogen (Pringle and Scheffer 1964). These substances are toxic to suscepts, but are essentially non-toxic to other living things. The known host-specific toxins produce all the symptoms of the disease caused by the respective pathogens and would be useful in studying host-parasite interactions (Strobel 1974).

Toxin was isolated from the culture filtrate of *X. campestris* pv. *oryzae* and *Drechslera oryzae*, the causal agents of bacterial leaf blight and brown leaf spot diseases of rice, respectively. The toxins from these pathogens had inhibited the growth of root of rice seedlings (Purushothaman and Prasad 1972, Lindberg 1971). *X. campestris* toxin was reported to be phenolic in nature.

Toxin-producing ability was qualitatively determined by one gene in *Cochlibolus carbonum* and *C. victoriae* (Scheffer et al. 1967). Lim and A.L. Hooker (1971) also observed that both selective pathogenicity and specific pathotoxin production of *C. heterostrophus* are monogenic in inheritance. The amount of pathotoxin produced may be polygenic (Scheffer et al. 1967). Similar inheritance of pathotoxin production was reported by Lim and A.L. Hooker (1971).

A cross between race T and O of *D. maydis* showed that one or more than one gene controls T-type pathogenicity and race T-type pathogenicity was associated with the ability to produce T toxin (Yoder and Gracen 1975). The pathogenicity expressed and amount of toxin produced are highly associated (Lim and A.L. Hooker 1971).

The toxin produced by race T of *D. maydis* is specific in inhibiting the elongation of primary roots of seedlings with Tcms cytoplasm. Primary roots of seedlings with normal cytoplasm are not so greatly inhibited (Lim et al. 1971). A host-specific toxin isolated from culture filtrates of race T of *D. maydis* to corn cytoplasm (cms-T) was designated as Hm-T toxin and a non-specific toxin to corn cytoplasm produced by race O was designated Hm-O toxin (Lim and A.L. Hooker 1972a). Both toxins are dialyzable, thermostable and give a weakly positive reaction with ninhydrin (Lim and A.L. Hooker 1972a). Extracts of corn leaves having cms-T cytoplasm infected with *D. maydis* race T exhibited a differential toxicity to resistant (normal) and susceptible (cms-T cytoplasm) corn similar to that of the pathotoxin (HmT-toxin) produced by the fungus in vitro and similar to the fungus itself (Lim and A.L. Hooker 1972b).

Toxin filtrate was obtained from 20- to 24-day-old cultures of *Drechslera sacchari*, the causal agent of eye spot disease of sugarcane. Toxin production and fungus growth were affected by both temperature and time. Symptoms produced by the toxin and the fungus are similar. Only plants susceptible to the fungus are affected by the toxin. The toxin reaction of 182 sugarcane clones was significantly correlated (r=0.88) to their reaction to the pathogen. Large-scale screening for resistance to eye spot disease can be accomplished accurately and rapidly by using this toxin (Steiner and Byther 1971).

Susceptibility or resistance to *D. victoriae* toxins appears to be expressed in passive and metabolically active tissues of oats (Samaddar and Scheffer 1970).

The role of host-selective toxin in colonization of corn leaves by *D. victoriae* and *C. carbonum* was studied by Yoder and Scheffer (1969) and Comstock and Scheffer (1973), respectively. They concluded that: (1) HC toxin is required for colonization of susceptible corn tissue by *Cochliobolus carbonum*; (2) dead or seriously damaged cells are not required for successful colonization (disruptive effects were not evident for more than 20 h after inoculation) of both the fungi; (3) inhibitory compounds produced by corn cells do not account for resistance to *D. carbonum* or to homologous pathogens; and (4) toxin produced from invading fungus, *D. victoriae*, causes the initial changes in the host cell physiology characteristic of disease development.

Inbred corn lines that were low, intermediate or high in resistance to *C. carbonum* had the same relative rank in sensitivity to HC toxin. Of several plant species tested (barley, corn, cucumber, oat, radish, sorghum, tomato and wheat), susceptible corn was most sensitive to HC toxin at a particular concentration, while resistant corn was most tolerant. The sensitivity of several non-host plants fell between these extremes. In contrast, a preparation of *D. victoriae* toxin completely inhibited root growth of susceptible oats at 0.009 µg/ml, but caused only partial inhibition of resistant oats and other plants (tomato, corn, wheat, sorghum and barley) at 3600 µg/ml (Kuo et al. 1970).

Yoder (1980) reviewed the literature on toxins in pathogenesis and concluded that the practical significance of pathologically important toxins is that they can act as reliable surrogates for the pathogens that produce them. This greatly simplifies bio-chemical analysis by permitting elimination of the pathogen from the system. It also facilitates screening for resistance among populations of whole plants, population of cells or protoplasts. Even organelles such as mitochondria or chloroplasts can be ac-curately identified in vitro or in vivo with appropriate toxins. Host-specific toxins could be of great use as a tool to select for disease-resistant cells or tissues in vitro.

4.14 Population Genetics of Gene-for-Gene Interactions

Van der Plank (1963, 1968) used the terms "directional" and "stabilizing selection" to systems of gene-for-gene relationships between major gene(s) of virulence and for resis-tance. Directional selection refers to selection favouring genes for virulence, i.e. selec-tion imposed by use of a resistant host cultivar. Directional selection is another name for adaptation. If a resistance gene is introduced to produce a new cultivar and if the pathogen adapts itself to the new situation by mutating to virulence or by increasing the frequency of virulence genes already present in the population that is directional selection towards virulence (Vanderplank 1978). Stabilizing selection was used by Vanderplank, as selection against unneccessary genes for virulence on susceptible hosts. Stabilizing selection, also called homeostasis, is the opposite of directional selection. It is resistance to change (Vanderplank 1978). Leonard (1977), however, noted that Vanderplank's definitions were a distortion of the concepts as developed in population genetics. The term stabilizing selection was originally applied to quantitative traits and was defined as the combination of opposing selection pressures directed towards the adapted mean of the population (P. Crill 1977). Stabilizing selection operates against extremes at either end of the scale for a trait. When the concept was extended to oligo-genic qualitative traits, stabilizing selection was again represented as the combination of opposing forces of selection (Mettler and Gregg 1969). In this case, the opposing forces of selection maintain the alleles at a given locus at stable equilibrium frequency in the population. The stable equilibrium provides optimum fitness for the population in that environment.

Vanderplank reasoned that gene-for-gene systems would never have been discovered if selection inevitably favoured resistance in the host population and virulence in the pathogen population. A gene for resistance cannot be recognized except in comparison with its allele for susceptibility, and in addition it will not be recognized if the corres-ponding gene for virulence is missing from the pathogen population. Oligogenic vertical resistance could not have been found unless genes for avirulence were sometimes selec-tively favoured over genes for virulence.

R.R. Nelson (1972) and P. Crill (1977) did not appreciate the arguments against Vanderplank's concept of selection against unnecessary genes for virulence because these arguments do not directly dispute the logic of Vanderplank's concept, many of them were rather based on examples and observations which misinterpreted the concept (Leonard and Czochor 1980). For instance, in competition studies a pathogen isolate with more identified genes for virulence is often found to be a stronger competitor than an isolate with fewer identified genes for virulence. This observation is not contrary to Vanderplank's concept, it is not even relevant to it.

The consequences of an absence of selection against unnecessary genes for virulence should be used to test the logic of Vanderplank's argument. The development of a gene-for-gene system may begin with a mutation for resistance. Because the plants with this gene are resistant, they produce more offspring on the average than do susceptible plants. As the gene for resistance increases in frequency, the ability of the pathogen to reproduce is reduced. If a gene for virulence arises by mutation and allows the pathogen to overcome the resistance, that gene for virulence will be favoured by selection (Person 1959); the host and pathogen populations, however, do not change instantaneously from susceptible to resistant and from avirulent to virulent. As long as the frequency of the gene for resistance is very low, selection for virulence occurs very slowly (Leonard 1977). The alleles for resistance and susceptibility and for virulence and avirulence will coexist in the host and pathogen populations, respectively, for a certain length of time. This type of polymorphism is known as a transient polymorphism. The population is polymorphic for the length of time required for one allele to replace the other. In balanced polymorphism, both alleles are maintained in the population by balancing or stabilizing selection (Leonard and Czochor 1980).

Vanderplank's concept is that gene-for-gene systems evolved as balanced polymorphisms. In agricultural systems the balance has been upset because the genes for oligogenic vertical resistance were used in pure line cultivars. Growth of these pure line cultivars over large areas caused the selective advantage to be tipped strongly in favour of the virulent races, and the formerly balanced polymorphism was converted into a short-lived transient polymorphism. In general, observations support that the gene-for-gene systems arose as balanced polymorphisms (Leonard and Czochor 1980). Except in cultivated crops grown as pure line cultivars, the greatest diversity in resistance genotypes and virulence genotypes in host and pathogen populations is usually found in areas where the pathogen is capable of causing more severe disease on susceptible hosts. Van der Plank (1963, 1968) summarized the evidence that pathogen races occur in greatest diversity in areas where genes for resistance in the host are most diverse.

In a race-specific gene-for-gene host: pathogen interrelation, genes of resistance are identified by the genes of virulence and vice versa. Separated genes of resistance can thus be used to analyze the virulence pattern of a parasite population and to understand the strategy developed by the pathogen. Using mainly the *Avena: Puccinia graminis* system and the different conditions in the United States and in Sweden, it is shown that different qualifications for reproduction of the parasite may change the adaptability pattern of virulence from a balanced polymorphism to an unnecessary accumulation (MacKey 1977).

Pathogen populations are more sensitive to changes in weather; the frequencies of genes for virulence would be expected to fluctuate much more widely from year to year than would the frequencies of genes for resistance (Leonard 1984). In general, genes for resistance in common gene-for-gene interactions typical of rusts, smuts, powdery mildews, and downy mildews would be expected to occur at low to moderate frequencies in co-evolving host populations, but in most cases the frequencies of genes for virulence would occur at high frequencies. The reverse would be true for gene-for-gene interactions involving host-specific toxins because genes for production of the host-specific toxins occur in very low frequencies, except in agricultural situations in which the toxin-sensitive host genotypes have been inadvertently increased to high frequencies in commercial crop production (Leonard 1977, 1978).

The virulence analysis developed by Wolfe and Schwarzbach (1975) and Wolfe et al. (1976) was applied to rice blast fungus population by Kiyosawa (1984). He studied the yearly change of the frequencies of virulence genes, $Av-i^+$ and $Av-k^+$, and genotype $Av-i^+$ $Av-k^+$ during a period from 1959 to 1965 in ten districts in Japan. Regression coefficients of frequencies of these genes showed negative values in all districts, except for Hokkaido, with statistically significant deviation of the coefficient from zero at the 1 or 5% level in some districts. This suggested that stabilizing selection is functional.

These observed frequencies of $Av-i^+$ $Av-k^+$ genotypes were higher than the expected frequencies under the assumption of independent occurrence of the two virulence genes in almost all districts and almost all years. Particularly high frequencies of $Av-i^+$ $Av-k^+$ genotypes in the former half-period cannot be interpreted by frequencies of resistance genes, Pi-i and Pi-k, because there are no or few genes in this period.

In addition to gradual increase or decrease of a special race during several years, sudden changes were observed in some districts (Kiyosawa 1984). These changes cannot always be explained by resistance gene frequencies.

Apart from these, many cases of non-random association of two or three virulence genes were illustrated. These deviations were not always explained by directional selection and/or stabilizing selection (Kiyosawa 1984). The seasonal change of fungus genotype frequency was observed in a few cases. Of these, one case showed similar seasonal change of fungus genotype frequencies. The genotype, $Av-a^+$ $Av-i^+$ $Av-k$, did pre-dominate in an early season and the genotype, $Av-a^+$ $Av-i$ $Av-k$, predominated in a late season on two varieties, Norin 6 ($Pi-a^+$ $Pi-i^+$ $Pi-k^+$) and Asahi 2 ($Pi-a^+$ $Pi-i^+$ $Pi-k^+$) in Saijo, Hiroshima Prefectures. If this change is not the result of chance, it may be due to selection in different directions in different seasons (Kiyosawa 1984).

The instances in which the change in genotype frequency cannot be explained by directional and/or stabilizing selection would be due to: (1) sampling error, (2) genetic random drift, (3) the influence of environment including invasion of fungus strains from outside source, and/or (4) inherent properties in host varieties, and/or pathogen strains.

The knowledge of the stability of gene frequency equilibrium conditions would be useful to plant breeders (Leonard 1984). This will help to understand the role of major genes to resistance in protecting plants in naturally co-evolving populations and in designing better strategies to obtain maximum benefits from those genes in agriculture. If such genes are used in cultivated crops at frequencies below their equilibrium frequencies in nature each major gene could provide modest but extremely long-lasting protection.

4.15 Fitness Models of Gene-for-Gene Interactions

Several plant pathologists have developed models of genetic interactions in gene-for-gene systems. Mode (1958) was the first to theoretically analyze genetic interactions between populations of plants and their pathogens.

Leonard and Czochor (1980) reviewed the different models and concluded that balanced polymorphisms in these systems are not possible without selection against unnecessary genes for virulence. They pointed that all except one of the models indicated that the equilibrium point for host and pathogen gene frequencies is locally unstable, and that the use of multiline variety with components at constant frequencies cannot be expected to lead to balanced, genetically diverse pathogen populations. Mode's third model (1961) is the only one in which the selection coefficients for the pathogen are assumed to be variable so that they may change with changes in pathogen gene frequencies and population densities. In all other models, selection is assumed to be hard selection that depends solely on differences in inherent rates of reproduction of different genotypes, rather than soft selection that can vary in intensity depending on competitive interactions among genotypes (B: Wallace 1975).

If in pathogen populations selection is primarily hard selection, the parameters in Leonard's (1977) model can be used to predict the fitness of pathogen populations selected by the use of susceptible cultivars, multiline varieties as advocated by Browning and Frey (1969) or pure line cultivars with pyramided genes for vertical resistance as suggested by R.R. Nelson (1972, 1979). The superior strategy should give lowest pathogen fitness. In the multiline variety, the fitness of the avirulent race would be less than 1 and the fitness of the virulent race would be between 1-k and 1-k+a (Table 4.9).

Table 4.9. Fitness of pathogen and host genotypes in a theoretical model by Leonard (1977)

	Fitness of pathogen genotypes in indicated combinations	
	rr (susceptible)	R – resistant
v (avirulent)	1	$1 - t$
V (virulent)	$1 - k$	$1 - k + a$
	Fitness of host genotypes in indicated combinations	
	rr (susceptible)	R – resistant
v (avirulent)	$1 - s$	$1 - c - s(1 - t)$
V (virulent)	$1 - s(1 - k)$	$1 - c - s(1 - k + a)$

k = cost of virulence; t = effectiveness of resistance in suppressing pathogen reproduction; a = advantage to the pathogen of having its virulence gene match the corresponding resistance gene in the host; s = suitability of the environment to disease development expressed in terms of loss of fitness of the susceptible host when attacked by the avirulent pathogen genotype; c = cost of resistance.

For a $>$ k, the multiline variety could be composed so that it would be superior to both the susceptible cultivar and the pure line resistant cultivar which would select the virulent race with a fitness of 1- k+a. Leonard and Czochor (1980) observed that in such situation, after the virulent race had been selected, the resistant cultivar would be inferior to the susceptible cultivar. For k $>$ a $>$ 0, the multiline variety would be superior to the susceptible cultivar. With hard selection, the pure line cultivar with pyramided genes for variety only if a $<$ 0. This is not likely to be a general situation and would preclude a balanced polymorphism in natural systems.

If selection in pathogen populations is, at least, partly soft selection that depends on competitive interactions, multiline varieties are even more likely to be superior to pure line cultivars with pyramided genes for resistance. With soft selection, gene-for-gene systems could have stable equilibrium points (Mode 1961), rather than limit cycles, and it might be possible to use multiline varieties to stabilize pathogen population with a diversity of races in stable equilibria. However, if selection coefficients in pathogen populations depend on the intensity of competitive interactions, it is likely that selection against unnecessary genes for virulence will be more intense when pathogen population densities are high than when they are low.

MacKenzie (1978) analyzed the fitness of benomyl-tolerant and benomyl-sensitive strains of *Cercospora beticola* using the data of Dovas et al. (1976). He indicated that selection depended almost completely on different inherent rates of reproduction of the strains rather than on competitive interactions between them. Changes in frequencies of tolerant and sensitive strains in mixtures in field plots could be predicted almost exactly from their individual rates of increase in separate plots. Thus selection was hard selection.

Ogle et al. (1973) found that changes in frequencies of pathogen isolates in mixtures on wheat plants for stem rust in the greenhouse could be predicted reasonably well from the fitness of each isolate by itself on separate plants. However, they observed that the rate of change in isolate frequencies in mixtures was slightly greater than that predicted from separate tests of the reproductive capacity of each strain. It appears that soft selection played a part in the change of isolate frequencies.

All of the evidence indicated that multiline varieties can provide adequate protection from disease if a sufficient number of genes for resistance are available. The number of components in the multiline variety can be kept at a manageable level by developing component lines with combinations of genes rather than a single gene for resistance. If this is done, the combinations should be planned on the phenotype-phenotype basis analyses (Wolfe et al. 1976) to detect combinations of virulence genes that may be less compatible than others. Certain combinations of virulence genes may be rare in the pathogen population because of linkage disequilibria or epistatic effects that reduce fitness in the pathogen. The disease control of the multiline might also be enhanced by systemic seed treatments rotated among the component lines in successive years in order to avoid or delay selection for fungicide tolerance in the pathogen population (Wolfe and Barratt 1977).

4.16 Vertical and Horizontal Resistance in Relation to Loss of Fitness

Vertical resistance in host dictates which races of the pathogen survive. If the pathogen does not possess genes required to incite disease in the host population, negative selection pressure occurs against non-pathogenic portion of the population (P. Crill 1977).

When a VR gene is present in the host, the pathogen must be able to develop races with virulence to match this gene if it is to survive (Van der Plank 1968). When the VR gene is absent from the host, and a matching virulence in pathogen is not necessary, then stabilizing selection will operate in favour of races of pathogen that have the fewest genes for virulence. The effectiveness of stabilizing selection in the pathogen population is determined by the relative strength of the VR gene in the host. The strength of a VR gene is in turn defined in terms of strength with which stabilizing selection acts against the complementary pathogen race. The stronger the gene for VR, the stronger the pressure for stabilizing selection.

Van der Plank (1968) states: "Underlying HR is a stabilizing selection operating against extremes: a genetic homeostasis. The stabilizing selection differs in origin from, but is as real as, the stabilizing selection on which VR rests. Stabilizing selection in the pathogen population when HR is used in the host is attributed to the stability of the races of the pathogen. This stability does not preclude the appearance of new races or the disappearance of old races; but maintains a stable balance among all the various races" (Van der Plank 1968).

Stabilizing selection in favour of the pathogen races without unnecessary virulence is considered to be the force behind VR. The races of the pathogen with excess virulence genes are less fit to survive and will therefore disappear from the pathogen population with time. This is due to the fact that the pathogen is forced to develop a substitute metabolic pathway for each additional virulence gene. The degree of inferiority of the substitute metabolic pathway employed by the pathogen to overcome the VR of the host gene that determines the strength of the resistance gene (Vanderplank 1975b). VR must be more ephemeral and therefore inferior to HR, which is more stable and longer lasting.

Stabilizing selection in pathogen races without unnecessary virulence is considered also to be the force behind HR. The mechanism of action is not the same as for VR, but rather is ascribed to the polygenic inheritance of aggressiveness (Van der Plank 1968). During each crop season that pathogen population will be exposed to a uniformly susceptible, but horizontally resistant variety and the survivors of the pathogen population will be all from the mean of the population. The intermediates of the population survive and reproduce while the extremes are lost. As Van der Plank (1968) states, races that survive are those with intermediate aggressiveness, not the extremely aggressive or unaggressive.

The cultivars developed in the absence of disease tend to have lowered HR. This is the evidence of the vertifolia effect. Plant breeders making selections among host populations protected from the disease by VR inadvertently tended to select progenies with low HR. By implication, this means that the optimal fitness in a population not threatened by disease does not coincide with the optimal fitness in a threatened population (Vanderplank 1978).

Vanderplank noted that this is also the evidence of many great epidemics. Populations of plants that have been grown for generations unthreatened by a pathogen often succumb badly to the pathogen when they meet it for the first time or after a long period of separation. A frequent feature of these epidemics, particularly among annuals or clonal host plants, is how quickly the epidemic decreases, largely due to the new selection pressure of the pathogen. For instance, when the potato virus Y^n (tobacco veinal necrosis virus) spread through the potato fields of central Europe in the late 1950's, there was a sufficient diversity of potato cultivars and advanced lines in breeders fields to absorb the shock within a few years; favourable old cultivars like Ackersegen were replaced by others much less susceptible, and, although spread of the virus cannot easily be controlled by insecticides, the potato industry has suffered little harm in the long term (Vanderplank 1978).

From the foregoing discussion it can be concluded that there is a price to pay, in lost fitness, for HR. Vanderplank noted that resistance itself is a gain of fitness in another direction, and a new balance of optimal fitness in the environment of disease is presumably struck. The loss of fitness of resistant hosts relative to susceptible hosts in the absence of the pathogen cannot be assessed. However, gene recombination may reduce the loss. For instance, the increased resistance to *Puccinia polysora* quickly developed by open pollinated maize in Africa could have been due to an accumulation of homozygotes (Van der Plank 1963), in which case gene recombination might in the course of time have reduced the loss of heterosis.

There is no equal evidence that VR, too, may bring with it a loss of fitness where disease is absent. In the absence of a pathogen and selection pressure from it, alleles for vertical susceptibility are common and those for resistance are rare in the host populations, which suggests a fitness advantage in susceptibility (Vanderplank 1978). Multilines grown from a mixture of seed of lines each differing by a resistance gene provide a counter to excessive loss of fitness as a result of increased resistance when this resistance is vertical, but not when it is horizontal.

4.17 Epidemiology of Vertical and Horizontal Resistance

The first essential of epidemiology is that it concerns populations, not individuals. If the individual is the unit of a pattern, then the epidemic is the pattern itself, the system (Kranz 1974). The second essential of epidemiology is that it concerns not one population but two, the host and the parasite, and above all the interaction of these two populations (R.A. Robinson 1976).

Vertical resistance reduces the effective amount of initial inoculum from which the epidemic starts and thereby delays the observed start. A feature of VR is that the infection rate is as fast as in the complete susceptible variety after the initial infection has occurred (Van der Plank 1968). Most VR is apparently controlled by single genes (R.A. Robinson 1971). Van der Plank (1968) assessed VR as follows: the effects of VR are strong in seasons of little disease when resistance is not very important, but weaker in seasons of much disease when resistance is most needed. Greater disappointments with VR have been in seasons of unusually high infection rate. Thus VR is portrayed as the basis for a "boom and bust" cycle of variety production.

The action of horizontal resistance is to slow down the epidemic after it has started (Van der Plank 1968). HR reduces the infection rate (so too does non-specific VR). HR can vary from zero or no resistance at all to practically absolute immunity. It includes klenducity, field resistance, and all those factors considered for disease tolerance (Schafer 1971). The effect of reducing the infection rate by HR is cumulative and greatest when the epidemic is of long duration. Four factors of resistance condition a low infection. These are: a lower proportion of spores that manage to initiate a lesion, a lower average spore production by a lesion, a shorter period of infectiousness, and a longer period of latency. For all of these factors examples are available in the literature.

One may use protectant fungicides to reduce infection rate or it may be done by controlling environment. One may reduce initial inoculum by sowing healthy vegetative propagating material, by crop rotation to reduce the abundance of the pathogens in the soil, by isolating fields from sources of infection, by fumigating soil etc.

The resistant varieties may be used to reduce both, i.e. either the initial inoculum or the rate of infection. VR commonly reduces initial effective inoculum, i.e. inoculum of races able to infect the crop. VR gains from diversity. If the pathogen is an obligate parasite, there must be at least two host varieties, whereas HR gains in potency if the pathogen enters the field from another field with HR. It will reach the field later and in smaller amounts because of the other field's HR, and it will develop in the field more slowly because of the field's own HR. HR gains from uniformity (Van der Plank 1968). VR is effective only if the pathogen enters the field from a field without VR or at least from a field with a different host genotype and different gene for VR.

If only one field in the whole area has HR, the effect of this resistance is at its lowest, while VR would have just reverse effect. The different response of HR and VR to diversity is one of the reasons why plant breeders have chosen VR. The plant breeder develops varieties which on leaving the nursery are resistant, because there are no prevalent races that can attack. The VR delays the epidemic for such a long time that no disease develops in the new variety. If the variety is agronomically superior it may occupy 100% of the acreage and at this time VR fades away (Van der Plank 1968). The VR is at its best in breeders' experimental plots, because of the diversity provided by the other varieties, which may also have VR. The variety with HR, on the other hand, is at its worst in breeders' nurseries in experimental trial plots. It is at its proper level of effectiveness in the large area of successful varieties. But if planted on a small scale, its HR will be less fully effective and may eventually be rejected by the farmers. Under such circumstances, multiline varieties will retain the special advantage.

5 Sources of Resistance and Methods of Testing for Resistance

Genetic variation is the prerequisite in any breeding programme. In case of breeding resistant varieties to diseases and insect pests, it is imperative to search for sources of resistance, i.e. the donors from which the resistant gene(s) may be transferred. The first step, therefore, is to collect variability including the wild/weedy relatives and land races of that species in which improvement in resistance to disease(s) and/or insect(s) is desired. Once the variability is collected, the next step is to screen the available gene pool against the important parasites. The resistant lines of immediate use may be included in the core germplasm which is used in hybridization. The rest of the lines may be kept in cold storage after proper classification. These germplasm lines may be grown in alternate years to revitalize their viability, etc. These may be screened against new parasites as and when they become important.

In this chapter, an attempt is made to describe the sources of resistance (both wild and cultivated types) to important parasites in some of the crop plants. The importance of germplasm in resistance breeding is also discussed by citing some of the important problems and achievements. In addition, the screening techniques for some of the destructive pathogens and insects are described. These instances of use of a gene pool and their screening may serve as a model for other parasites.

5.1 Sources of Resistance

The supply of genes for resistance, for disease(s), insect pest(s) and nematode(s) is the first concern in an on-going resistance breeding programme. The primary and secondary gene centres of cultivated plants are the best places to find genuine resistance to common diseases and pests. This rule first formulated by Vavilov (1922, 1957) was restated by Zhukovskij (1959, 1961). The information on the origin and evolution of cultivated plants and their wild progenitors has been summarized by Vavilov (1949–1950) and Zhukovskij (1964). Zhukovskij (1970) suggested "micro-centres" — the restricted geographical areas with primary gene centres of several endemic species. Several micro-centres in a larger area are united into a "mega-centre". He also used the term "geno-fund" to indicate the genetic source or gene pools for breeding. The centres of origin are found in every continent, except in Australia (Harlan 1975b).

A reverse method has been proposed by which an attempt is made to locate the gene centres of cultivated crops according to their host-specific pests and pathogens (Leppik 1966). He showed that certain specialized parasites and their distribution on

Table 5.1. Germplasm at international centres. (Swaminathan 1980)

Crop	Number of accessions	Centre
Rice	40,000	IRRI
Rice	4500	IITA
Maize	13,000	CIMMYT
Sorghum	17,000	ICRISAT
Pearl millet	11,900	ICRISAT
Minor millet	·2400	ICRISAT
Wheat	24,000	ICARDA
Barley	13,000	ICRISAT
Chickpea	11,300	ICRISAT
Pigeonpea	6000	ICARDA
Chickpea	3300	ICARDA
Grondnut	7700	ICRISAT
Cowpea	8000	ITTA
Lentil	6000	ICARDA
Broad bean	1800	ICARDA
Soybean	800	IITA
Bambara groundnut	3400	IITA
Lima bean	1200	IITA
Phaseolus sp.	21,000	CIAT
Potato	12,000	CIP
Cassava	2500	CIAT
Cassava	3050	IITA
Forages	4500	CIAT
Forages	7300	ICARDA

particular plant groups can serve as reliable indicators that help to trace the origin and evolution of their host (Leppik 1968).

The International Board of Plant Genetic Resources (IBPGR) was established in 1973, has sponsored plant collections in over 20 countries or areas. All of the major cereals, the pulses, grasses, forage legumes, groundnuts, vegetables, potatoes, cassava and sweet potatoes have been included. A large collections of many species are maintained at the international centres distributed around the world (Table 5.1). In addition, in some countries several research centres maintain germplasm collections (Sprague 1980).

a) Use of Land Races/Cultivated Types. The resistance most directly useful in plant breeding is found in cultivars of the same species. One of the earlier experiments conducted to show that variation for resistance exists was by Barrus (1911). He did not find any variety of *Phaseolus vulgaris* that was resistant to every strain of the bean anthracnose fungus *(Colletotrichum lindemuthianum)*. However, some varieties were less susceptible than others. Since then in every important crop species the variation for resistance is reported against almost all important diseases and insect pests and in most cases the gene(s) have been transferred. The reports in literature are infinite on this subject and it is not possible to cite them all; however, some of the cases are listed in Tables 5.2 and 5.3. In most of the cases the gene(s) have been incorporated from the

Table 5.2. Sources of resistance reported in cultivated types for some of the crop plants against important plant pathogens

Crop	Disease	Reference
Rice	All major diseases	Khush (1977a)
	Blast *(Pyricularia oryzae)*	S.V.S. Shastry et al. (1971)
	Blight *(Xanthomonas campestris* pv. *oryzae)*	S.V.S. Shastry et al. (1971)
	Rice tungro virus (RTV)	S.V.S. Shastry et al. (1971), D.P. Singh and Nanda (1980)
	Bacterial leaf streak *(X. campestris* pv. *translucens)*	Nanda et al. (1977)
Wheat	All three rusts	Bahadur et al. (1979), G.S. Sharma and R.B. Singh (1975), Kochumadhavan et al. (1980)
	Powdery mildew *(Erysiphe graminis)*	Bahadur et al. (1979)
	Leaf blight *(Drechslera sativa)*	Joshi and Adlakha (1974)
Barley	Powdery mildew *(E. graminis)*	Upadhyay and Prakash (1980)
	Loose smut *(Ustilago nuda)*	P.L. Thomas and Metcalfe (1984)
Pearl millet	Green ear disease *(Sclerospora graminicola)*	Chahal et al. (1975)
Maize	Downy mildews *Sclerophthora* and *Peronosclerospora* spp.	Lal and Singh (1985)
	Stalkrots, Erwinia, Cephalosporium and *Fusarium* spp.	Lal and I.S. Singh (1985)
	Leaf blight *(Drechslera turcica)*	Aujla and Chahal (1975)
Sugarcane	Red rot *(Glomerella tucumanensis)*	Gill et al. (1984)
Cowpea	Aphid-borne mosaic virus	Patel et al. (1982a,b)
	Bacterial blight *(X. campestris* pv. *vignicola)*	Patel (1983)
Urd bean and Mungbean	All major diseases	D.P. Singh (1981)
	Mungbean yellow mosaic virus (MYMV)	D.P. Singh (1982)
Soybean	Mungbean yellow mosaic virus (MYMV)	B.B. Singh et al. (1974a,b)
	Rust *(Phakopsora pachyrhizi)*	B.B. Singh et al. (1974a)
Frenchbean	Anthracnose *(Colletotrichum lindemuthianum)*	Schwartz et al. (1982)
	Angular leaf spot *(Isariopsis griseola)*	
	Halo blight *(Pseudomonas phaseolicola)* *P. syringae*	M.L. Schuster and Coyne (1981)
Cucurbits	All major diseases	Sitterly (1972)
Musk melon	Gummy stem blight *(Mycosphaerella melonis)* *M. citrullina*	Sowell (1981)
Apple	Apple scab fungus *(Venturia inaequalis)*	E.B. Williams and Kuc (1969)
Sunflower	*Phytophthora drechsleri* *P. parasitica* *P. cryptogea*	Da Via et al. (1981)

Table 5.3. Sources of resistance reported in cultivated types for some of the crop plants against important insect pests

Crop	Insect pest	Reference
Rice	All major insect pests	Khush (1977a)
	Brown planthopper, BPH *(Nilaparvata lugens)*	Krishna et al. (1977)
	Gall midge *(Pachydiplosis oryzae)*	S.V.S. Shastry et al. (1971)
	Yellow stemborer *(Tryporyza incertulas)*	S.V.S. Shastry et al. (1971)
	Green leafhopper *(Nephotettix* spp.)	S.V.S. Shastry et al. (1971), D.P. Singh and Nanda (1980)
Soybean	Lima bean podborer *(Eticlla zinckenella)*	Talekar and Chen (1983)
	Southern green stik *(Nizara viridula)*	Gilman et al. (1982)
	Mexican bean beetle *(Epilachna varivestris)*	Hallman et al. (1977)
	Corn ear worm *(Heliothis zeae)*	C.M. Smith and Brim (1979)
	Cyst nematode *(Heterodera glycines)*	Anand and Gallo (1984)
Soybean and Mungbean	Bean flies *(Ophiomgia phaseoli) (O. centrosematis) (Melanogromyza sozae)*	H.S. Chiang and Talekar (1980)
	Mexican bean beetle *(E. varivestris)*	John et al. (1971)
Urd bean and Mungbean	Whitefly *(Bemisia tabaci)*	D.P. Singh (1982)
Peanuts	Two-spotted spider mite *(Tetranychus urticeae)*	D.R. Johnson et al. (1982)
Corn	Maize stalk borer *(Chilo partellus)*	Panwar et al. (1979)
	Black cutworm *(Agrotis ypsilon)*	R.L. Wilson et al. (1983)
Jute	Mite *(Memitarsonemus latus)*	Dash and Dikshit (1982)

land races or derivatives of land races. In general more reports are on the successful transfer of disease than of insect pest resistance.

It is worth citing the case of rice varietal improvement work. The collections of rice assembled through exploration in the Himalayan foot hills of north-eastern India yielded a large number of primitive rice cultivars with resistance to major diseases and insect pest, including blast, bacterial blight, tungro virus, gall midge and stemborers, and some with combined resistance (S.V.S. Shastry et al. 1971). A systematic screening for diseases and insect pest resistance was conducted at IRRI, the International Rice Research Institute, the Philippines (T.T. Chang et al. 1975). Some of the rice germplasm lines have been reported to have resistance to more than one biotype and also to insect pest (U.P. Rao and Seetharaman 1980).

The sources of resistance against a particular parasite in general are obtained from the centres of origin, where pressure of parasite is also very high. The only known source of resistance to barley yellow dwarf virus (BYDV), a virus disease of small-grained cereals which at times causes serious damage, was discovered in 1963 among Ethiopian accessions in the USDA barley collections (Qualset 1975). Khandelwal and Nath (1979) reported that resistance to fruitfly *(Dacus cucurbitae)* in water melon *(Citrullus lanatus)* is possibly obtained from the area of secondary centre of origin of the crop and the area of highest insect pest infestation. Similar is the case of tropical grain legumes grown in the Indian sub-continent. However, in other cases the presence of resistance alleles appears to be unrelated to the presence of the pest. Resistance to the hessian fly has been found in some wheat accessions in Portugal, although the pest is unknown in that area. Collections of *Oryza glaberrima* rice were highly resistant to the green leafhopper, *Nephotettix virescens*, although this pest is not known to occur in Africa. Many Asian varieties of *O. sativa* were highly resistant to delphacid, *Sogatodes oryzicola*, which is only found in the Americas (Pathak and R.C. Saxena 1980).

The donors for resistance used in breeding programmes are mostly resistant to one race or biotype of parasite, i.e. race-specific resistance has been exploited. The resistant gene(s) become ineffective as soon as a more virulent race or biotype originates. Therefore when new and additional collections are to be made, they ought to be screened and used in the on-going breeding programmes. The search for germplasm lines with multiple resistance may be continued. Such work is being conducted in rice at IRRI, the Philippines and in cowpea at IITA, Nigeria. There is scope for screening the germplasm against general resistance or the horizontal kind of resistance, which may be comparatively more durable.

b) Use of Wild/Alien Species. Sometimes adequate resistance does not appear to exist in cultivated species and then the breeder has no alternative but to search for resistance in related species or genera. Often the genes from wild species/genera have resistance against a wide range of races. Such genes have been called "super genes". Therefore, resistance from wild relative is attractive even when other sources of resistance are available. I.A. Watson (1970b) noted that the new races/biotypes of parasite overcome the resistant gene(s) being used in the cultivars. Thus it is emphasized that the wild relatives or species will become an increasingly important source of germplasm in the breeding of many crops. Harlan (1976) observed that the wild races or species have been exploited most often as sources of disease, insect and nematode resistance in almost every crop grown by man. The ancestors and other relatives of crop species have acquired a wealth of resistance gene(s) to parasites over long periods of host-pathogen co-evolution (Frankel 1977).

In the transfer of gene(s) in wide hybridization, the major obstacle is in making crosses; when the two species are distantly related, fertilization either fails to occur or embryo development stops at an early stage or embryo may abort. Reciprocal differences are not uncommon. The slow pollen tube growth in interspecific hybrids has been overcome by stylar grafts. Growth hormone/embryo culture may be used to check the embryo abortion and to obtain hybrid plants. The chromosome number of wild species may be doubled to facilitate crossing with wild species at a higher level. Hybrid plants may at times die or are sterile. This can be improved by colchicine treatments of hybrids to produce amphidiploids (Knott and Dvorak 1976). In general while using wide crosses, repeated back-crosses to the recurrent parent and selection are required to recover all the characteristics.

Some of the reports on the wild/alien species used to transfer resistant gene(s) for diseases or which can be used in future as parents are presented in Table 5.4. Some of these cases are discussed here in detail. The wild species of cucumber viz; *Cucumis harwickii*, *C. leptodermis* and *C. myriocarpus* have multiple resistance to several common diseases and insect pests (Leppik 1966). Similarly *Triticum timopheevi* and *T. monococcum* are resistant to all types of rust disease of wheat. The wild species has been

Table 5.4. Some of the wild/related species of different crops used to transfer or possessing genes for disease resistance

Crop	Disease/pathogen	Wild/related species	Reference
Wheat	Stem rust or black rust (*Puccinia graminis* f.sp. *tritici*)	*Triticum boeticum*	Gerechter-Amitai et al. (1971)
		T. monococcum	Kerber and Dyck (1973)
		T. durum	I.A. Watson and Stewart (1956)
		T. timopheevi	Nyquist (1962)
		Agropyron elongatum	Knott (1961, 1964), F.C. Elliot (1967)
	Brown rust or leaf rust (*P. recondita* f.sp. *tritici*)	*A. intermedium*	Wienhues (1966)
		Aegilops umbellulata	Sears (1956)
		A. speltoides	Dvorak (1977)
		A. squarrosa	Kerber and Dyck (1969)
		Agropyron elongatum	D. Sharma and Knott (1966)
		A. monococcum	D. Sharma and Knott (1966)
		A. intermedium	Wienhues (1966)
		Secale cereale	Driscoll and Jensen (1964)
	Yellow rust or stripe rust (*P. striiformis*)	*Aegilops comosa*	Riley et al. (1968)
	Rust (all types)	*T. timopheevi*	Harlan and Zohary (1966, Briggle and Vogel (1968, Zohary et al. (1969)
		T. monococcum var. *hornemannii*	Harlan and Zohary (1966)
		T. militinae mutant from *T. timopheevii*	Leppik (1970)
	Eye spot (*Pseudocercosporella herpotrichoides*)	*Aegilops ventricosa*	Kimber (1967)
	Powdery mildew (*Erysiphe graminis* f.sp. *tritici*)	*Secale cereale*	Driscoll and Jensen (1963)
	Barley yellow dwarf and wheat streak mosaic virus	*Agropyron* species	H.C. Sharma et al. (1984)
Barley	Powdery mildew (*E. graminis* f.sp. *hordei*)	*Hordeum spontaneum*	Biffen (1907)
		H. bulbosum	I.T. Jones and Pickering (1978)
		H. laevigatum	Moore (1977)
Pearl millet	Rust (*Puccinia substriata* var. *indica*)	*Pennisetum americanum* subspecies *mondii*	Hanno et al. (1985)
Sweet sorghum	Rough leaf spot (*Ascochyta sorghina*)	*Sorghum* species	Zummo and Broadhead (1984)
Oats	Stem rust or crown rust (*P. graminis* f.sp. *avenae*) or *P. coronata*	*Avena sterilis*	Browning and Frey (1959), Martens et al. (1968, 1981), Simons (1985)

Table 5.4 (continued)

Crop	Disease/pathogen	Wild/related species	Reference
Rice	Grassy stunt virus	Oryza nivara	Khush and Ling (1974)
	Stem rot (Sclerotium oryzae-sativae)	O. rufipogon	Figoni et al. (1983)
		O. nivara	
		O. spontanea	
Sugarbeet	Cercospora leaf spot (Cercospora beticola)	Beta procumbens	Coons (1954)
Groundnut	Peanut mottle virus	Arachis diogoi	Melouk et al. (1984)
		Arachis species	
	Curly top virus	Solanum pennelli	M.V. Martin (1962)
		L. peruvianum	M.V. Martin and Thomas (1969)
		L. pimpinellifolium	
	Spotted wilt caused by virus	L. pimpinellifolium	Finlay (1953)
		L. peruvianum	Hutton and Peak (1949)
Grapes	Downy mildew (Plasmopara viticola)	Vitis unifera var. sylvestris	Zhukovskij (1965)
Tomato	Wilt (Fusarium oxysporum f.sp. lycopersici)	Lycopersicon pimpinellifolium	Bohn and Tucker (1939)
		L. hirsutum	Kerr and D.L. Bailey (1964)
	Leaf mould (Cladosporium fulvum)	L. peruvianum	Leppik (1970)
	Many diseases	L. hirsutum	Skrdla et al. (1968)
	Blight (Septoria lycopersici)	L. pimpinellifolium	Lincoln and Cummins (1949)
	Bacterial canker (Corynebacterium michiganense)	L. hirsutum	Thyr (1968)
		L. hirsutum	
	Tomato foot and stem rot (Didymella lycopersici)	L. pimpinellifolium	Boukema (1982)
		L. hirsutum glabratum	
	Tobacco mosaic virus	L. hirsutum	F.O. Holmes (1938, 1939)
		L. chilense	
		L. hirsutum	
Tobacco	Leaf spot (Cercospora nicotianae)	Nicotiana repanda	Pelham (1966)
	Early blight (Alternaria solani)		Stavely et al. (1973)
	Black root rot (Thielavia basicola)	N. debneyi	Clayton (1958)
	Black shank (Phytophthora parasitica var. nicotianae)	N. plumbaginifolia	Apple (1962)
	Wild fire (Pseudomonas syringae pv. tabaci)	N. longiflora	Clayton (1947b)

Crop	Disease	Species	References
	Black fire (P. angulata)	N. longiflora	Clayton (1947a)
	Angular leaf spot (P. angulata)	N. longiflora	Clayton (1958)
	Tobacco mosaic virus	N. glutinosa	F.O. Holmes (1938), Valleau (1952)
Cotton	Bacterial blight (X. campestris pv. malvacearum)	Gossypium arboreum	R.L. Knight (1954)
	Boll rot (Gibberella fujikuroi)	G. tomentosum	Meyer and Meyer (1961)
Soybean	Yellow mosaic virus	Glycine formosana	B.B. Singh et al. (1974b)
	Bud blight	Glycine soja	Orellana (1981)
Mungbean	Mungbean yellow mosaic virus	Vigna radiata var. sublobata	D.P. Singh and B.L. Sharma (1983)
Peppermint	Wilt (Verticillium sp.)	Mentha cripsa	Murray (1961)
	Rust (Puccinia menthae)	Mentha arvensis	D.D. Roberts (1981)
		M. citrata	
		M. aquatica	
		M. rotundifolia	
Sunflower	Rust (Puccinia helianthi)	Helianthus agrophyllus	Putt (1964), Sackston (1956, 1957)
	Downy mildew (Plasmopora halstedii)	H. petiolaris	
French bean	Common blight (Xanthomonas campestris pv. phaseoli and X. campestris pv. phaseoli var. fuscans)	Phaseolus acutifolius	M.L. Schuster and Coyne (1981)
		P. coccineus	
Chickpea	Blight (Ascochyta rabiei)	Cicer pinnatifidum	K.B. Singh et al. (1981)
		C. montbretti	
		C. judaicum	
Pigeonpea	Stem blight (Phytophthora drechsleri)	Atylosia species	Kannaiyan et al. (1981)
Potato	Wart (Synchytrium endobioticum)	S. commersonii	Hawkes (1958)
		S. tartijense	
		S. chacoense	
		S. vereni	
		S. curtitibolium	
		S. maglia	
		S. juzepeczuki	
	Late blight (Phytophthora infestans)	Solanum demissum	Saleman (1931), Bukasov (1936), Rudorf (1950), Ross (1966)

Table 5.4 (continued)

Crop	Disease/pathogen	Wild/related species	Reference
		S. bulbocastanum	Leppik (1970)
		S. dimissum	Niederhauser and Mills (1953)
		S. cardiophyllum	Niederhauser et al. (1954)
	Common scab (Streptomyces scabies)	S. commersonii	Reddick (1939)
		S. chacoense	
	Powdery mildew	S. kurtzianum	Mastenbrock (1955)
	Charcoal rot (Macrophomina phaseolina)	S. chacoense	Paharia et al. (1962)
	Black leg (Erwinia sp.)	S. horovitzii	Dobias (1978)
		S. subtilus	
		S. boergeri	
	Brown rot (Pseudomonas solanacearum)	S. chacoense	Thung (1947)
	Virus X	S. cartilobum	Cockerham (1943)
		S. juzpeczuki	
		S. chacoense	
	Virus A	S. cartilobum	Cockerham (1943)
		S. chacoense	
		S. infundibuliforme	
Brassica	Clubroot (Plasmodiophora brassicae)	Brassica campestris	Lammerink (1970), Johnston (1974)
		Brassica napus	M.S. Chiang and Crete (1983)

used as parent more frequently in the genus *Lycopersicon* (Skrdla et al. 1968). They are reported to have resistance to bacterial wilt, bacterial canker, *Fusarium* wilt, grey leaf spot, leaf mould, *Septoria* leaf spot, *Verticillium* wilt, curly top virus, mosaic virus, three kinds of spotted wilt virus, root knot nematode, and more. These resistances have been transferred to commercial cultivars (Rick 1967). *L. pimpinellifolium* has contributed more genes for resistance to diseases than any other tomato species. It is estimated that at least 30 commercial varieties have this species in their parentage and there are probably more (Skrdla et al. 1968). The only known source of resistance to grassy stunt virus in rice at the present time is *Oryza nivara*. It was found after screening thousands of cultivated rices and hundreds of wild ones. Resistance was found in only one of the accessions and not all of the plants in that population had the desired gene (Khush and Ling 1974). The resistance is transferred to several high-yielding lines/varieties of rice.

The wild relatives of barley, *Hordeum vulgare*, were screened using double inoculation with pathogenic cultures for resistance to *Erysiphe graminis* f.sp. *hordei*, the causal agent of powdery mildew (Moseman et al. (1981). PI 282660, PI 284750, PI 354937 and PI 296836 had one gene; PI 296836 had three genes; and PI 296800 had four genes against more virulent cultures. This shows that the wild species have a wider spectrum of resistance against the plant pathogens.

At Pantnagar University in India *Vigna radiata* var. *sublobata* and *V. mungo* var. *silvestris* the wild progenitors of mungbean, *V. radiata* Wilczek and Urd bean, *V. mungo* (Fig. 5.1) have been used for increasing variation for yield components and for incorporating resistance to mungbean yellow mosaic virus (MYMV).

Numerous reports are available in the literature on the utility of wild relatives for insect resistance Table 5.5. The wild species of the genus *Helianthus* have been evaluated more systematically for insect pest resistance. C.E. Rogers and Thompson (1978b) evaluated the germplasm from different species of sunflower. They evaluated it in the laboratory for resistance to carrot beetle, *Bothynus gibbosus*. Significantly less (50% level) root injury by the carrot beetle occurred with *H. argophyllus*, *H. ciliaris*, *H. grosseserratus*, *H. laetiflorus*, *H. maximilianni*, *H. mollis*, *H. niveus canescens*, *H. scalicifolius* and *H. tuberosus* than with the commercial Hybrid 896. Resistance in wild *Helianthus* to the sunflower beetle *(Zygogramma exclamationis)* (Rogers and Thompson 1978a) was also screened in laboratory feeding trials. All the larvae died after feeding on *H. ciliaris* and *H. salicifolius*. Also significantly more larvae died after feeding on wild *H. tuberosus* than after feeding on Hybrid 896. The total development period for larvae reared on the wild annual sunflowers, and consumption of leaf area by larvae reared on *H. petiolaris*, *H. tuberosus* and *H. paradoxus* were significantly below that of larvae reared on Hybrid 896. Females feeding on the wild species of *Helianthus* did not lay eggs.

c) **Multiplasm as a Source of Disease Resistance.** The outbreak on maize of T race of *Drechslera maydis* often in epiphytotic proportions has focussed attention on the importance of cytoplasms in plant disease control. It was due to widespread use of single cytoplasm, i.e. T cytoplasm in maize. The multiplasm proposal suggested by Grogan (1971) is designed especially as a safeguard against a mono-cytoplasm, particularly one as a male sterile, and consequent disease problem especially in cross-pollinated crops

Fig. 5.1. *Vigna mungo* var. *silvestris* (narrow leaves) and *V. radiata* var. *sublobata* (broad leaves), the wild progenitors to Urd bean and Mungbean respectively found in the vicinity of Pantnagar University in India

such as maize and sorghum. The component lines for use in the female parents should be converted to two or more cytoplasms with common restorers. The stocks could be maintained separately and mixed equally in hybrid production, or by using a mixture of equal portions of each cytoplasm and reproduced with the normal maintainer counter part.

d) Induced Mutants as Source of Resistance. Plant breeding involves genetic variability and utilization of induced variation in selection. Genes and gene forms or alleles altered by mutations and recombination constitute the genetic variation. Recombination is the result of meiotic or mitotic crossing-over or of independent assortment. The resultant genotypes are selected either by natural or artificial selection.

Table 5.5. Some of the wild/related species of different crops used to transfer or possesses genes for insect-pests resistance

Crop	Insect pest	Wild/related species	Reference
Wheat	Greebug (Schizaphis graminum)	Triticum durum	Starks and Burton (1977)
	Cereal leaf beetle (Oulema melanopus)	T. turgidum	Leisle (1974)
	Cereal cyst nematode (Heterodera avenae)	T. durum	Sikora et al. (1972)
Barley	Cereal cyst nematode (Heterodera avenae)	Hordeum pallidum	Nilsson-Ehle (1920), Fiddian and Kimber (1964)
	Stem eelworm (Ditylenchus dipsaci)		
Oats	Oat stem eelworm (D. dipsaci)	Avena byzantina	Griffiths et al. (1957)
		A. ludioviciana	
	Root knot nematode (Meloidogyne spp.)	A. strigosa	S. Anderson (1961)
		A. abyssinicia	
		A. sterilis	
Cotton	Boll worm (Heliothis zeae)	Gossypium barbadense	Cook (1906)
	Boll weevil (Anthonomus grandis)	G. armourianum	Meyer (1957)
Potato	Golden cyst nematode (Globodera rostochiensis) (G. pallida)	Solanum vereni	Dunnett (1960)
	White potato cyst nematode (G. pallida)	S. multidissectum	Fullar and Howard (1974)
		S. tuberosum ssp. andigena	Dale and Philips (1982)
	Globodera sp.	S. vernei	Ellenby (1954), Mai and Peterson (1952), Ross (1966), Dunnett (1957), Fullar and Howard (1974)
		S. balsii	
		S. sucrense	
		S. aplocense	
		S. matinae	
		S. canasense	
	Leafhopper (Empoasca fabae)	S. pampasense	Klashorst and Tingey (1979)
	Aphids (Myzus persicae)	S. chacoense	J.B. Adams (1946)
	(M. euphorbiae)	S. bulbocastanum	Radcliffe and Lauer (1970)
		S. michoacanum	
		S. steriophyllidum	
	Thrips	S. berthaulti	Gibson (1979)
		S. tarijense	
Tobacco	Root knot nematode (Meloidogyne javanica)	Nicotiana repanda	Stavely et al. (1973), Burk (1967)
Tomato	Root cyst nematode (Heterodera marioni)	Lycopersicon peruvianum	Hare (1965), McFarlane et al. (1946)

Table 5.5 (continued)

Crop	Insect pest	Wild/related species	Reference
Sugarbeet	Cyst nematode (*Heterodera schachtii*)	*Beta patellares*	Savitsky and Price (1965), Heÿbroek
		B. procumbens	et al. (1983)
Cucumber	Root knot nematode (*Meloidogyne incognita*)	*Cucumis anguria*	Fassuliotus (1970)
		C. ficifolius	
Raspberry	Fruit fly (*Dacus cucurbitae*)	*Momordica charantia*	Pal et al. (1984)
	Aphids (*Amphorophora agathonica*)	*Rubus odoratus*	Keep (1972)
		R. coreanus	
		R. occidentalis	
Lettuce	Leaf aphid (*Nasonovia ribis nigri*)	*Lactuca virosa*	Eenink et al. (1982b)
Clover	Stem nematodes (*Ditylenchus dipsaci*)	*Ladino clover*	W.M. Williams (1972)
Alfalfa	Spotted alfalfa aphid (*Therioaphis maculata*)	*Medicago scutellata*	Pandey et al. (1984)
		M. rugosa	
		M. tittoralis	
Sunflower	Potato leaf hopper (*Empoasca abrupta*)	*Helianthus species*	C.E. Rogers (1981)

The mutation which is the basic ingredient of plant breeding and evolution can be broadly described as "all genetic alternations, ranging from single base substitutions within the genetic material, the deoxy-ribose nucleic acid (DNA) comprising the gene to gross changes in chromosome number and structure which cannot be accounted for by recombination" (Nilan et al. 1977). The mutations could be either spontaneous or induced. Through spontaneous mutations only relatively few genetic changes occur and therefore only a limited number of new genetic recombinants are produced and this in turn leads to a process of slow evolution. To achieve the rapid evolution needed by plant breeders, more rapid genetic forms are required. Such requirements can be fulfilled only with the aid of induced mutations as supplements to natural variability.

High effective mutagenic agents both physical and chemical are available which can be used for creating variation. Factors like genotype, nuclear volume of cells, temperature, pH, metabolic activity, oxygen and water content of the irradiated cells, and the stage of cell cycle influence effectiveness and efficiency of mutation induction in plants for physical mutagens. Similarly for chemical mutagens, the physical and chemical properties of the mutagen such as solubility, hydrolysis half-life, and reactivity, concentration of mutagen, duration of treatment, temperature of the treatment solution and treated cells, pH of the solution, stages of cell cycle, and genotype and metabolic activity of the cells affect the effectiveness and efficiency of mutagen (Nilan et al. 1977).

Resistance has been induced for powdery mildew in barley (Jorgensen 1971); in wheat to bunt (Bozzini 1971); and *Septoria nodorum* (Little 1971), in peppermint to wilt, *Verticillium* (Murray 1969), and in soybean to rust, *Phakopsora pachyrhizi* (Tsai et al. 1974). Swarup and Raghava (1974) induced resistance to leaf curl virus in *Zinnia elegans* after irradiation with 20 kilorad X-ray.

The rice variety Norin 8 was susceptible to all races of blast fungus in Japan. In the treated generations, mutant lines with improved resistant reactions to blast disease were selected and their frequency was about 0.1% of the M_2 strains (Yamasaki and Kawai 1968). The resistant mutant of rice blast fungus was also reported in Korea from the leading *japonica* variety, Palkweng (Ree 1971). In Thailand many mutants for better blast and gall midge resistance were reported from Nahng-Mon, S–4, Khao-Tah-Haeng 17, Nangh-Phaya 132, and Muey-Nawng 62N, respectively (Khambanonda 1971). Two mutant lines have been selected from IR 8 with better resistance for bacterial leaf blight disease in Ceylon (Gunawardera et al. 1971). In India, mutants selected for *indica* grain type had better resistance to bacterial leaf blight than the *japonica* mother variety (Micke et al. 1971). Induced mutation for blast resistance in a high-yielding variety of rice were reported by Kaur et al. (1977) and for bacterial leaf blight in the variety Vijaya were reported by Padmanabhan et al. (1977) through the use of the EMS mutagen.

Attempts have also been made to induce variation for resistance to insect pests. In two varieties of *Brassica juncea*, Laha 101 and RL 18, treated with different concentrations of the mutagens, ethylmethane sulphonate (EMS), maleic hydrazide (MH), ethylene imine (EI), diethyl sulphate (DES) and different doses of gamma rays; several plants highly resistant to the attacks of the aphid, *Lipaphis pseudobrassicae* were isolated in the M_2 generation. The two varieties responded differently to the different mutagens, EMS and gamma rays inducing more aphid resistant mutations in Laha 101 and HA and EI in RL 18 (Srinivasachar and Verma 1971). As aphid resistance may not be a case of simple inheritance, effective control over the pest could be achieved by concentrating the resistant genes in a single strain.

e) Methods of Testing for Resistance to Diseases. Inoculation techniques differ from disease to disease; however, an optimum level of infection may be obtained. There are a number of devices by which it is possible to control the level of infection in greenhouse inoculations. Some of these are adjustment of temperature, humidity, age of the plant, concentration of inoculum and uniformity of application. In the field, however, the breeder depends on the natural conditions and he will be fortunate to obtain records at a uniform and optimum level of infection. The field records are many times confused by the presence of a multiplicity of diseases and irregular plant development. It is, therefore, imperative that disease tests are conducted under conditions of controlled inoculation. In the case of screening large number of germplasm lines, it will be desirable to screen them in field conditions, and the promising ones ought to be screened in the glasshouse under controlled conditions. Methods of inoculation have been developed for most of the important diseases; however, they are subjet to further improvement. Some breeders/pathologists have found that they can save time, space and materials by inoculating the same plant with more than one pathogen either simultaneously or in succession, but in view of the evidence for antagonism among micro-organisms it is generally inadvisable to mix pathogens on the same plant. For the same reason it is inadvisable to rely on field observations solely if controlled inoculations can possibly be made, because it is only rarely that only one pathogen is present in a particular field (Andrus 1953).

Since there are a large number of diseases in a crop species and there are numerous important cultivated species, it is not possible to describe the screening techniques for all the diseases in different crops. An attempt, therefore, has been made here to list the methods of screening of some of the important diseases of important crops.

5.2 Screening for Resistance to Diseases

5.2.1 Fungal Diseases

a) Cereal Rusts. The seedling inoculation method is useful in screening types with vertical resistance. The field assessment of adult plant or horizontal resistance is carried out in rust nurseries (Z. Kiraly 1971). Cultivars to be tested are sown in small plots in the nursery. Usually a series of plots are planted in a row between border rows or "spreader rows" of a susceptible cultivar. A few plants in the spreader are inoculated in the boot by a hypodermic injection. This provides abundant inoculum for the experimental plots between the spreaders. The spreader rows can also be inoculated with a suspension of uredospores in a 0.1% agar, solution.

Lelley (1957) proposed the "circle plot" method. By this method, in the middle of a circle of the wheat cultivar or mutant to be investigated, two to three plants of a very susceptible spreader are inoculated hypodermically at the boot stage. From these plants the inoculum is easily distributed to the surrounding plants in the circle.

The most reliable method of infecting cereals with rusts is the hypodermic injection of individual plants. However, this is a time-consuming procedure (A.H. Kiraly 1971).

Detached leaf culture technique can also be followed in the laboratory. Using this method, one can inoculate different leaf segments from a plant with different races and even with different leaf pathogens without infecting the whole plant.

b) **Phytophthora infestans.** Resistance based on hypersensitivity or VR can be screened in potted plants of *Solanum*. The leaves of the plants are inoculated with zoospores of the fungus and plants are incubated at 18 °C in a moist chamber for 24 h. Detached leaves can also be used for inoculation. The leaves are put in Petri dishes and one drop of zoospore suspension is put on the under surface of each leaflet. These are incubated at 18 °C for 12 h. Necrotic spots usually appear on the 3rd day after inoculation.

The field resistance or HR can be evaluated in the laboratory or by peroxidase test (Kiraly 1971). Laboratory assessment of field resistance is carried out on leaf discs inoculated with a drop of zoospore suspension. Inoculated discs are kept on moist filter paper at 15 °C and assayed after about a week. The leaves for such evaluation are taken from 2-month-old plants (Hodgson 1961) and collected from one third of the distance from top to base. Good correlation is exhibited between results in the field and in the laboratory.

Field resistance can be assessed in field trials after the appearance of the first symptoms of the disease. If four replicates are used, then 1000 leaflets per cultivar may be assessed (5 leaflets per leaf × 5 leaves per stem × 10 stems per plot × 4 replicates per cultivar). The number of leaflets destroyed is expressed as percentage of the total. Weekly observations are necessary to follow the development of the disease (Lapwood 1961).

The peroxidase test is based on the relation of activity of peroxidase in leaves and field resistance of potato to late blight (Kiraly et al. 1970). Sakai and Toriyama (1964) have shown a good correlation between this enzyme and resistance in the adult stage but is uncertain in the young stages.

c) **Wilt Diseases.** Wilts caused by *Fusarium* and *Verticillium* species constitute an important problem in many crops. Artificial inoculation for screening-resistant types is carried out by dipping roots in a suspension of spores or propagules. In this method seedlings are immersed in the suspension of spores before transplanting (Wright 1966). Strawberry seedlings at the third leaf stage, i.e. after about 45 days, were inoculated with pathogenic isolates of *Verticillium dahliae* by immersing the roots in a conidial suspension (10,000 ml) for 1 h, then replacing the seedlings in the liquid nutrient solution. Wilt symptoms, i.e. leaf wilting and necrosis, were first observed on the most susceptible seedlings 21 days after inoculation. All seedlings were re-inoculated 35 days after initial inoculation, and those which were virtually healthy 50 days later were graded as tolerant and subjected to further tests (Jordan 1973). The soil inoculation method was very effective in producing wilt when inocula aged 3 weeks or more were used. Age and vigour of tomato seedlings had no significant correlation with disease *(F. oxysporum)* severity. However, the disease progressed more rapidly on younger seedlings (Kesavan and Choodhury 1977).

Resistance to *Fusarium oxysporum* was detected using green house-screening techniques, and a rapid test tube screening method was developed that determined resistance levels in 16 days in soybean (S. Leath and Carroll 1982). Cultivars resistant to *F. oxysporum* were also detected, using a mycelium production index. This new method is not destructive and allows plants to be saved for use in a breeding programme.

d) Powdery Mildews. The seedling inoculation technique is useful in screening types with VR. In this method the young plants are dusted with conidia shaken from heavily infected plants. The field assessment of adult plant resistance is uncertain. Environmental conditions are not always congenial for the spread of disease. The detached-leaf culture method offers a possibility of evaluating resistance (Moseman and Greeley 1966).

e) Downy Mildews. The downy mildew disease of maize (caused by *Sclerophthora* and *Peronosclerospora* spp.) causes considerable yield losses in different parts of the world.

The powdered infected leaves of maize with brown stripe downy mildew, *Sclerophthora rayssiae* var. *zeae*, collected during the last season or from early plantings in the same season containing oospores, can serve as a source of inoculum. Freshly infected leaves of 2–3 cm pieces could be put in the whorl of the seedlings (Lal and I.S. Singh 1984). Scoring can be done using a 1–5 index, where 1 = no infection and 5 = very heavy infection (Fig. 5.2).

For sugarcane downy mildew (*Peronosclerospora* spp.), epiphytotic conditions are created by planting every fifth row of spreader 20 days prior to planting of test material. The spreader, as well as the test material, is inoculated from infected leaves collected at 01.00 h. The conidia are washed off in water to prepare suspension which is inoculated by spraying at 02.00 h at two leaf stage, followed by two to three inoculations on alternate days (Lal and I.S. Singh 1984).

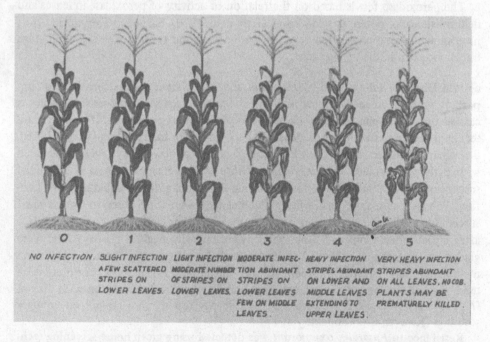

0	1	2	3	4	5
NO INFECTION.	SLIGHT INFECTION A FEW SCATTERED STRIPES ON LOWER LEAVES	LIGHT INFECTION MODERATE NUMBER OF STRIPES ON LOWER LEAVES.	MODERATE INFECTION ABUNDANT STRIPES ON LOWER LEAVES FEW ON MIDDLE LEAVES.	HEAVY INFECTION STRIPES ABUNDANT ON LOWER AND MIDDLE LEAVES EXTENDING TO UPPER LEAVES.	VERY HEAVY INFECTION STRIPES ABUNDANT ON ALL LEAVES, NO COB. PLANTS MAY BE PREMATURELY KILLED.

Fig. 5.2. The rating scale for brown stripe downy mildew (with the courtesy of I.S. Singh)

The sorghum downy mildew (SDM) caused by *Peronosclerospora sorghi* cannot be grown artificially, therefore the conidial inoculum must be produced on diseased plants. Craig (1976) gave a method of inoculating sorghum, *Sorghum bicolor*, with conidia of SDM. In this method air was used to scatter conidia over the plants. The efficiency in the separation of sorghum lines resistant and susceptible to SDM agreed with the reactions of these lines to the field inoculum.

An effective, large-scale field-screening technique to identify sources of resistance to downy mildew, *Sclerospora graminicola*, in pearl millet *(Pennisetum typhoideum)* was reported by R.J. Williams et al. (1981). The technique is based on preplanted infector rows that provide sporangial inoculum. It has been successfully used to identify and improve downy mildew resistance in composites, cultivars, and hybrid parents.

f) Smuts. Artificial inoculation against smut diseases are rather simple. The seeds are powdered with teliosphores in the case of covered smuts (*Tilletia* and *Ustilago* sp.) and these surface-contaminated seeds are sown in the field. Teliospore suspension can be injected into the flowers in cereals with loose smut *(Ustilago nuda)* or into tissues of other organs in corn with *U. zeae* and *Sphacelotheca reiliana* (Kiraly et al. 1970). An effective field-screening technique was developed to identify resistance to smut in pearl millet by Thakur et al. (1983a). It involved inoculation of tillers at the boot leaf stage by injecting a sporidial suspension of *Tolyposporium penicillariae* into the space around the inflorescence within the flag leaf sheath, followed by covering the tiller "boots" with parchment bags. High humidity was maintained with frequent overhead sprinkler irrigation throughout the period of inoculation, flowering and grain development. Inflorescences were scored for smut reaction 20-25 days after inoculation with the aid of standard key developed to estimate percent severity.

g) Bunts. A screening technique against karnal bunt *(Neovossia indica)* disease of wheat was reported by Aujla et al. (1980). The culture of the pathogen was established on yeastal-potato-dextrose on liquid as well as on solid medium in the presence of soil extract. The viable sporidial inoculum (1 ml) was inserted into the plant by hypodermic syringe at the boot leaf stage when the awns were just visible along the flag leaf. The inoculations were also done by clipping the central florets and applying the inoculum (1-2 drops) with the help of a syringe. The pots containing inoculated plants were incubated for a week in a muslin cloth chamber, where high humidity was provided by spraying water with a knapsack sprayer, where as in the biological incubators 90-100% humidity, 15°-18 °C temperature and 1600 lx (8-10 h) light were provided. In the field, inoculated plants were provided with 90-100% humidity using perfo-spray equipment (mist sprayer). The minimum and maximum temperatures during inoculations were 10° and 23.2 °C respectively.

h) Leaf Blight. A rapid method for selecting plants for control of corn leaf blight was suggested by Gracen et al. (1971). It consists of spraying entire flats of plants in the greenhouse or rows in the field to determine resistant and susceptible plants. One or more leaves of the plants can be injected to allow for the determination of resistant and susceptible plants. Plants 2 or 3 weeks old were sprayed or injected with toxin solutions to determine resistance.

Aujla and Chahal (1975) used the blighted debris with viable spores of *Drechslera turcica* to screen the maize inbred lines. The blighted debris were put in the leaf whorls at a uniform rate. The disease data were recorded on a 1 to 5 scale, where 1 = no disease and 5 = severe disease.

5.2.2 Bacterial Diseases

a) **Bacterial Leaf Blight**. A needle inoculation method for bacterial leaf blight (*Xanthomonas campestris* pv. *oryzae*) disease of rice *(Oryza sativa)* was proposed by Muko and Yoshida (1951). In this method an inoculation pad with six needles mounted on it in two rows was fixed on the thumb, while another pad with a cushion of cotton covered with cheese cloth was fixed on the middle finger. It has been suitably modified by IRRI pathologists. The pads were periodically soaked with inoculum and the flag leaves punctured mid-way between the base and the tip on either side of the mid-rib. At least three flag leaves were inoculated per plant. The disease scoring was done 20 days after inoculation using a 0-9 scale based on the extent of the lesion development from the point of inoculation. The disease score of 0-3, 4-6 and 7-9 are considered as resistant, moderate, and susceptible respectively. The parental reactions were taken into account while classifying the plants. At IRRI (1966) other methods of inoculation have also been tried. They include (a) cutting the roots of seedlings and then dipping for 2-3 min in bacterial suspension and then transplanting in pots; (b) cutting the upper portion of leaves and dipping in bacterial suspension; and (c) spraying bacterial suspension on seedlings. The root-cutting method induced Kresek symptoms most quickly. Leaf cutting also induced Kresek, but spraying was the slowest method. Kauffman et al. (1973) gave an improved method for inoculating rice with bacterial leaf blight. It consists of clipping off the tips of rice leaves with scissors whose blades were dipped in bacterial suspension containing 10^9 cells per ml. Five 60- to 80-day-old plants can be successively inoculated. Although the technique measures tissue susceptibility, good correlation exists between tissue susceptibility and natural infection. This method is more efficient.

The concentration of the bacterial suspension must be lower than 10^6 cells ml^{-1} if inoculation is by hypodermic injection or high-pressure spraying (Klement 1968). On the other hand, by using low-pressure sprays or simply a brush for inoculation, and thereby distributing the bacteria on the surface of the plant organs, the concentration can be higher than 10^6 cells ml^{-1}.

b) **Bacterial Pustule**. A simplified method for inoculation for bacterial pustule (*X. campestris* pv. *phaseoli*) of soybean was developed by J.P. Jones and Hartwig (1959). Freshly infected leaves were used as a source of inoculum by first running them through a food chopper and then commuting this material in a food blender with tap water. This suspension is allowed to stand for about 2 h before filtering and is then used to inoculate plants. The chopped infected leaves may be stored at low temperature for use in the following year.

c) **Stalk Rot**. This is caused by *Erwinia chrysanthemi*, and is an important disease of corn in tropical countries. The inoculum is increased on nutrient broth for 48 h at 30 °C and diluted ten times with sterile water to maintain concentration of $1 \times 10^{7-9}$ cells per ml. The inoculation is done by making a diagonal hole, deep to the pith, in the middle of second internode at pre-silking or until flowering. The plants in a plot may be rated using a 1-5 scale (where 1 = limited infection to very small spot in the pith at the site of inoculum and 5 = disease spread over three or more internodes, plants wilted and may have toppled down) or the wilted/healthy plants in a plot may be counted after 15 days of inoculation. For observation, plants are cut from the ground in such a way that the first basal internode is intact (Lal and I.S. Singh 1984).

5.2.3 Tests for Resistance Using Toxins

The development of in vitro techniques for the breeding or the vegetative propagation of plants using cultures of embryos, calluses, anthers, somatic cells or protoplasts, opens up new opportunities for the selection of individuals which are resistant or tolerant to plant diseases. Screening for resistance to some pathogens can be carried out in their absence, i.e. by using toxins produced by them which kill the tissues of the host plant. Toxins have been used as agents of selective pressure to screen oat mutants resistant to *Drechslera victoriae* (Wheeler and Luke 1955); of sugarcane resistant to *D. sacchari* (Byther and Steiner 1972) and of maize resistant to *D. maydis*, race T (Wheeler et al. 1971, Lim et al. 1971). The oat variety Victoria and its derivatives are susceptible to Victoria blight *(D. victoriae)*. The toxin produced by the fungus is called Victorin. It is highly specific to kill germinating seeds of Victoria oats. Toxin concentrations lethal to plants of the variety Victoria have no effect on plants of oat varieties resistant to *D. victoriae*. The Victorin screening procedure is one of the most efficient yet discovered for detecting mutants. Konzak (1956) used it for selecting resistant seedlings from seeds exposed to thermal neutrons and X-rays. The seeds harvested from individual M_1 plants were rolled in moistened paper towels, allowed to germinate in a moist chamber for about 2 days, then sprayed with diluted toxin solution. After toxin treatments, seeds are allowed to continue germination for another 3 or 4 days at 75° to 80 °F. The rolls were opened and resistant mutants transferred to soil. The toxin kills the roots from the sensitive plants, but does not affect these from resistant individuals in the same lot of seeds.

Schertz and Tai (1969) screened germinated sorghum seedlings for resistance to Milo disease caused by *Periconia circinata*. They immersed roots of seedlings in culture filtrate of *P. circinata* for 4 days. Surviving seedlings were transplanted and self-pollinated and their progeny tested. One in ten proved to be mutant; the rest were escapes.

Dutrecq (1977) developed a technique to test the inhibition of germination of barley seeds by culture filtrate of *D. sativa*. Seeds were soaked for 24 h in toxin preparations, and then placed for 48 h in wet sand. A quasilinear relationship was found between seed germination and toxin concentration. The toxic effect was markedly affected by pH of culture filtrates, being maximum at pH 2.0. Control pH 2.0 filtrates of uninoculated medium did not affect the germination of barley seeds under similar conditions.

Such tests are easier to carry out and they can cause the same kind of symptoms as do natural infections of the pathogens concerned.

5.3 Virus Diseases

a) **Transmission by Grafting.** Since viruses go systemic in their host, they also pass through grafts. Plant viruses can be inoculated through graft unions between diseased and healthy plants by allowing vascular union between stock and scion. It has been used for inoculating yellow vein mosaic virus in okra *(Abelmoschus esculentus)* by Varma (1952). Although grafting is very laborious and time-consuming, it has been used in large-scale programmes of selecting for resistance, for example in testing for field immunity of potato virus X, which involves hypersensitivity controlled by a single gene, Nx. The presence of resistance gene Nx in a potato plant can easily be recognized by grafting on to it a scion from a susceptible variety which is infected with potato virus X. Plants with the gene Nx show a top necrotic reaction 2 to 4 weeks later (Russell 1978).

Graft transmission is limited to plants that are closely related but, for instance, solanaceous species of tobacco, tomato, potato, and thornapple may be graft-compatible.

b) **Transmission with Dodder.** For species that do not graft successfully, dodder *(Cuscuta)* may be a last resort to establish vascular contacts between virus donor and acceptor plants (Bos 1983). Several species, particularly *C. campestris* and *C. subinclusa*, have been used to transmit viruses but more especially mycoplasmas. The dodder forms bridges between diseased and healthy plants.

c) **Mechanical Transmission.** Many viruses can be transmitted mechanically by rubbing on to the healthy plants with a suspension of virus particles. Purified virus preparation is preferable, although transmission can usually be obtained with crude or clarified sap from infected plants, therefore this transmission is referred to as sap inoculation (Russell 1978). The inoculum can be applied in various ways, for example by stroking the plants with a virus-contaminated finger, piece of muslin or soft brush, or by a spray gun which injects inoculum deeply into the tissues of the host plant. A spray gun has been used successfully to inoculate sugarbeet plants with curly top virus (Mumford 1972). Screening against leaf crinkle virus of urdbean is done by rubbing the sap at the primary leaf stage (Nene 1972). To transfer mosaic mottle virus in legumes, Nene (1972) seedlings are inoculated mechanically by rubbing the inoculum, prepared in a sodium phosphate buffer of pH 7.6. Mechanical transmission has also been used in cowpea for black eye mosaic virus; in red clover for vein mosaic virus, in potato for virus S and in watermelon and *Phaseolus vulgaris* for watermelon mosaic virus 1 and watermelon mosaic virus 2, respectively, and in several other cases. The inoculum can be made in phosphate buffer and in general primary leaf (first leaf stage) is inoculated. If required, a second inoculation may be followed after some interval.

d) **Transmission Through Vegetative Propagation.** Viruses as a rule infect host plants systemically. Hence, all plant parts including those used for vegetative propagation, such as bulbs, corms, roots, scions and budwood, are invaded by the virus (Bos 1983).

e) **Transmission Through Vector.** Mungbean yellow mosaic virus of grain legumes is transmitted through whitefly, *Bemisia tabaci* (Genn.) and causes considerable losses in crops like greengram, blackgram and soybean. The germplasm can be screened by planting spreader rows after every five to six rows of the test material (Fig. 5.3). No

Fig. 5.3. The spreader row screening technique used for screening of white-fly transmitted mung-bean yellow mosaic virus (MYMV) in mungbean and urd bean

insecticide should be sprayed during the crop season, in order not to disturb the population increase of the vector (D.P. Singh 1981). Nene (1972) suggested mass-screening and individual plant-screening techniques. In the mass-screening technique cages of 60 X 90 X 120 cm dimension covered with muslin cloth bags are used to cover the rows of test material. For screening individual plants of F_1 or parents, Nene suggested the use of insect-proof transparent plastic pickle pots with a screwcap. In both the methods for acquisition feeding, whiteflies are collected from field and are kept in starvation for a period of 3–4 h, and then fed on diseased plants for 20–24 h. The viruliferous flies were then transferred to other cages to feed on 20–40-day-old plants for inoculation feeding. Eight to ten viruliferous flies per plant were released and allowed to feed for a period of 3 to 4 h. The scoring of individual plants is done on a 1–9 scale after 3 weeks of inoculation.

A similar technique was suggested for screening resistance to grassy stunt virus (Ling et al. 1970) and tungro virus (Ling 1974) diseases of rice transmitted by *Nilaparvata lugens* and *Nephotettix virescens*, respectively. In these cases also, the vector is made viruliferous, after which they are used to inoculate seedlings.

Sterility mosaic of pigeonpea occurs widely in the Indian sub-continent. It is transmitted by the eriophid mite, *Aceria cajani*. Nene (1972) reported transmission of disease by tying the branches of diseased plants on to the healthy ones (the twig-tying technique). Later Nene and M.V. Reddy (1976) reported a more efficient technique, the "leaf-stapling technique". In this technique the leaflets from diseased plants with mites are stapled to the primary leaves of the test seedlings. Using this technique, 100% infection was obtained in 10–15-day-old seedlings of susceptible cultivar Sharda.

f) **Transmission Through Seed and Pollen.** In recent years, there have been many reports of virus transmission through seed. Over 100 viruses are known to be seed-borne in at least some of their hosts (Bos 1983). Each infected seed may produce a new source in the next season or later at another site. Pollen transmission may result in the production of virus-infected seed on healthy mother plants, as with alfalfa mosaic and bean common mosaic viruses, but with several other viruses infected pollen is less viable than virus-free pollen and cannot compete with it.

5.4 Sequential Inoculations for Multiple Disease Resistance

A breeding programme was carried out in cowpea, *Vigna unguiculata* to incorporate resistance to cowpea aphid-borne mosaic virus (CAMV), cowpea mosaic virus (CMV), bacterial blight (BB), and bacterial pustule (BP) diseases in otherwise high-yielding and widely adapted cultivars found to be promising in multilocation trials in Tanzania (Patel et al. 1982b). They followed sequential inoculation screening of the segregating populations in the field for multiple disease resistance. The rub inoculations in the susceptible and resistant plants with a mixed sap prepared from CPMV and CMV stock cultures did not show synergistic or inhibitory effects among the viruses. Each line reacted as it was expected to behave to individual viruses.

In screening against bacterial pustule (BP), the pathogen suspension was infiltrated in leaves with the help of a hypodermic syringe. This inoculation method was necessary to distinguish brown hypersensitive resistant (BHR) and non-hypersensitive resistant (R) type resistance which was not possible by spray inoculation.

Bacterial blight (BB) was rub-inoculated in the trifoliates and assessed on inoculated leaves, and the disease developed in other leaves due to secondary spread. Early inoculation provided ample time for splash dispersal, leading to severe blight in the susceptible plants.

When the progenies of the plants selected as resistant in the field were screened individually against CAMV, CMV, BP and BB in the pot house, it was found that sequential inoculation was fairly effective in screening susceptible and resistant segregants for multiple disease resistance.

Several screening techniques for plant pathogens have been described in detail. It is necessary to test the material in an on-going breeding programme in the field as well as in the glasshouse under controlled conditions. The field tests facilitate the testing of large populations, while the glasshouse tests are important because conditions before and after inoculation can be standardized and plants are exposed to uniform inoculum. Therefore, both tests are complementary and not a substitute for each other.

5.4.1 Assessment of Disease Resistance

Whatever method is used for classifying disease reaction, it should be remembered that an accurate assay of resistance depends on the level of infection. At higher levels of infection, the different grades of susceptibility tend to merge and disappear and, therefore, valuable differences are overlooked, while at a low level of infection too many plants fall erroneously into a resistant class. For every disease, there is an optimum level of infection at which various degrees of resistance are most clearly distinguishable, and only long experience will enable the breeder to know the optimum level. Parlevliet (1981) noted that disease symptoms can be assessed in various ways.

1. As the proportion of plant units infected, that is, percentage of diseased plants or ears.

2. As a proportion of the total area of the plant tissue affected by the disease (W.C. James 1974).

3. More detailed evaluations measure the number and the size of successful infections. For pathogens that spread according to the compound interest law (Van der Plank 1963), disease severity is the cummulative result of infection frequency (proportion of spores that result in sporulating lesions), latent period (time from infection to spore production), spore production (spores produced per lesion or per unit area of tissue per unit of time) and infection period (period of sporulation).

A different method of assessment is used with many biotrophic leaf pathogens, such as rust and powdery mildews, to which the host may react to give high resistance. With this reaction type, a halo of necrotic or chlorotic tissue develops around the infection point and pathogen growth is hindered qualitatively (no sporulation) or quantitatively (restricted sporulation). The disease development is rated by infection types (IT), where IT 0 equals a necrotic or chlorotic fleck without sporulation and IT 4 equals a sporulating pustule without chlorosis or necrosis. IT 1, 2, and 3 describe pustules surrounded by necrosis or chlorotic tissue with increasing intensities of sporulation. Generally IT 0, 1, and 2 are classified as resistant and IT 3 and 4 as susceptible. Resistance to low IT type is race-specific and may be expressed at seedling or adult plant stage only.

Remote-sensing techniques may provide subjective disease assessment by the use of microdensitometers and electronic scanners to quantify disease severity as exhibited on infrared (IR) aerial photographs. Such a technique has been used in potato crop against late blight disease. This allows whole plots or field to be scored, thereby removing the sampling error normally associated with conventional sampling procedures and disease assessment methods.

Loegering et al. (1976) developed a method of rating single plants for general resistance (GR). It directly compares, over time, the relative amount of disease on a single plant with all other plants in the test. The method is simple, rapid, and reproducible. It involves taking notes many times (7-14) throughout the disease cycle on the basis of S = standard for the most diseased, as observed from a standing position on each day notes are taken, and L = less than standard. The number of times during the season that a plant is rated S is the disease index.

5.5 Methods of Testing for Resistance to Insect Pests

According to Painter and Peters (1956), a screening technique must provide for a study of many strains in a minimum of time and space, a reasonably accurate classification of resistance with reproducibility of results, and an opportunity to discern the presence of various components of resistance: non-preference, antibiosis and tolerance. Plant resistance to insects can be described in terms of either the reaction of the insect or the reaction of the plant, or as either an effect or a result. The technique/design for resistance to insects vary with the insect or plant. However, it is important to examine a large quantity of diverse material at the initial stage based on the effect either on host or pest. The preferred procedure is to discard the obviously susceptible germ-plasm lines using mass-screening, field or greenhouse tests. The selected strains are then retested intensively to confirm their resistance and to discard the pseudo-resistant plants, including cases of escape and host evasion (Painter 1951). J.N. Jenkins (1981) suggested that the procedure for evaluating insect resistance must have a good supply of eggs, larvae or adult insects for infesting plants in a breeding nursery. If insects are reared in the laboratory, they must represent the wild population in vitality, biotype composition, and genetic structure, and they must be nourished so that their behaviour and population capabilities are similar to those in the wild.

In the foregoing pages, screening techniques for some of the important insect pests in the important crop species are described in detail.

a) **Stemborers.** Stemborers are serious pests of the rice crop. More than 1400 rice varieties have been screened for resistance to the striped borer *Chilo suppressalis* (Pathak and R.C. Sacena 1980). Varietal screening is done in paired rows, 5 m long, perhaps in un-replicated trial. Up to 3000 varieties can be accommodated. The planting is timed so that the maximum tillering coincides with the harvest of neighbouring rice crops; this helps in heavy infestations because moths emerging from the maturing crop migrate to oviposit on the younger plants.

The borer incidence is recorded as dead hearts 60 days after transplanting and as white heads near harvest. The percentage of damage is calculated as follows:

$$X = P \bar{x} nz \, 100$$

where,

$$P = \frac{\text{Number of affected hills}}{\text{Number of hills in the plots}}$$

$$\bar{x} nz = \frac{\text{Number of dead hearts}}{\text{Number of tillers in the affected hills}}$$

X = Percentage dead hearts or white heads .

The varieties with low infestation are tested in replicated trials following the same method. Resistant and susceptible checks are planted after every ten plots. The most resistants are tested in the greenhouse or laboratory by various tests using first-instar larvae.

b) Leafhoppers and Planthoppers. The green leafhoppers *(Nephotettix virescens)*, brown planthoppers *(Nilaparvata lugens)*, and white-baked planthopper *(Sogatella furcifera)* cause considerable damage to the rice crop. Athwal et al. (1971) and Pathak and R.C. Saxena (1980) suggested screening techniques; the "bulk screening test" and the "tiller test" for resistance to these pests.

The bulk screening test is more efficient and is used frequently. It consists of planting the test material in rows about 5 cm apart in wooden flats of $60 \times 45 \times 10$ cm dimension. Normally in a flat 12 to 13 rows are accommodated. While screening F_2 plants, parent varieties occupy half of the rows. To ensure that all seedlings are at the same stage of growth at insect infestation, the F_2 seeds and parental varieties may be first germinated in Petri-dishes and then planted in the wooden flats filled with soil. In the F_3 tests, the F_3 progenies and parental varieties may be seeded directly in rows about 20 cm long obtained by dividing the 45-cm row length at the centre. A single flat occupies 20-24 lines with about 25 seedlings in each line. One row of each of the susceptible and resistant parents is interspaced at random for each group of eight to ten rows of the test material. The greenhouse temperature is maintained between 25° to 30 °C. At one leaf stage the seedlings are infested with insects from virus-free insect colonies. Second and third instar nymphs are used to inoculate the test material. About seven or eight insects per seedling are enough to distinguish the resistant and susceptible types. Susceptible seedlings start yellowing and ultimately die in 1 to 2 weeks. When the seedlings of the susceptible parent are either dead or wilted, the reaction is noted on a 0-9 scale, as suggested by Pathak and R.C. Saxena (1980) (Table 5.6).

The varieties rated resistant or moderately resistant are further evaluated by determining the survival of adult or nymph insects on individual potted plants.

The tiller test consists of infesting individual plants of the F_2 and parents with 15 insects and classifying the reaction on the basis of insects survival. The single plants are grown in 15-cm pots and one tiller of 6-week-old plants is placed in a cage and 15 second and third instar nymphs are released into the cage. For the leafhopper test,

Table 5.6. Rating for damage by leafhoppers and planthoppers (Pathak and Saxena 1980)

Grade	Rating	Seedling damage by		
		Green leafhopper	Brown planthopper	White baked planthopper
0	Highly resistant	No damage	No damage	No damage
1	Resistant	First leaf yellow	First leaf partially yellow	First leaf yellow-orange
3	Moderately resistant	50-70% of all leaves yellow	First and second leaves partially yellow	50% of leaves or their tips yellow-orange, slight stunting
5	Moderately susceptible	All leaves yellow	Marked yellowing, stunting	Most leaves or their tips yellow-orange, stunting
7	Susceptible	50% of plants dead	Severe wilting and stunting	50% of plants dead, severe wilting and stunting
9	Highly susceptible	Plants dead	Plants dead	Plants dead

a cylindrical cage of nylar plastic, 53 × 8 cm in diameter is used to enclose the entire tiller. For the brown planthopper test, a smaller cage 13 × 4 cm in diameter is used to enclose a stem of a tiller. The cages have nylon mesh ventilators. The living insects on individual plants are counted about 10 days after infestation. Most insects on resistant plants die within 10 days, while on susceptible plants normal growth and development is observed.

c) **Aphids.** An attached vs. excised trifoliate technique was used for the evaluation of resistance in alfalfa to the spotted alfalfa aphid, *Therioaphis maculata*. The adult and nymphal survival was used to compare the level of resistance between caged trifoliates on the plant and the corresponding plant part which had been detached. The results suggested that using excised trifoliate procedure in resistance screening tended to under-estimate the resistance level of the plant population tested (J.G. Thomas et al. 1966). Uninjured Cody alfalfa plants were selected in the field under a heavy infestation of pea aphid, *Acyrthosiphon pisum* by Harvey et al. (1972). Resistance appeared compar-able to that obtained by selection in the greenhouse. Progeny testing facilitated selec-tion of highly resistant plants.

Kennedy and Schaefers (1974) evaluated immunity in the cuttings of cultivars with varying resistance in red raspberry to aphid, *Amphorophora agathonica*, via cuttings. They reported that the method is rapid and requires less time and space than does the use of potted plants.

An excised leaflet test was developed to assess green-peach aphid, *Myzus persicae* resistance of wild tuber-bearing *Solanum* species germplasm by Sams et al. (1975). Correlations of excised leaflet tests with field evaluations were significant at a 1% level of probability. Genotype-environment interaction was low in field evaluations. Pre-liminary selection of important clones by the excised leaf test followed by final selec-tion after a replicated field evaluation is suggested.

The mustard aphid, *Lipaphis erysimi*, is a pest of the genus *Brassica*. The various criteria used in screening include the extent of aphid population during a given period, development, fecundity and longevity of the aphids and the yield of the cultivars (Pathak 1961 and several others). The percentage of seedling survival was used as criterian by Jarvis (1970). Bakhetia and Bindra (1977) reported that *Brassica* germ-plasm may be screened against aphid at any stage, provided the optimum level of the aphid population is used. However, the seedling stage may be preferred owing to its ease in handling, and also it requires less time and aphid population. In addition, systems are obtained early and material is tested at uniform bio-climatic conditions in insectory.

d) **Mites.** A method of seedling screening suitable for mass evaluation of cotton germ-plasm (M.F. Schuster et al. 1973) involves mite cultures maintained on bush beans under conditions of light, temperature and humidity. Seeds of cotton lines to be evalu-ated are planted in the greenhouse, and 5-day-old seedlings are infested by placing bean leaves on top of the small plants. Mites move from bean leaves on to cotton seedlings. The plants are rated for resistance 24 to 45 days after using a damage scale of 1 to 5. An injury index for each line is calculated by multiplying the score by number of plants in each category and dividing by total number of plants for that line.

Ponti and Inggamer (1976) gave an improved leaf disc technique for biotests in the mites. The technique enables the efficient screening of large populations of plants under controlled conditions. The production of individual mites (two-spotted spider mite, *Tetranychus urticae*) in cucumber was observed by placing single female mites on leaf discs. The discs are punched from the leaves and placed upside down on filter paper. The filter paper lies on a sheet of 3-mm-thick foam plastic, which floats in a plastic tray on a watery solution of benzimidazole (10 mg^{-1} l) which retards the senescence.

e) Bollworm. R.L. Wilson and F.D. Wilson (1975) compared an X-ray and a green boll technique for screening cotton for resistance to pink bollworm, *Pectinophora gossypiella*. When an X-ray method of examining the damage done by bollworm to cotton seed was compared with an examination of green bolls for number of insects, the data obtained by both methods were significantly correlated. The main advantages of the X-ray method are that it provides a direct measure of plant (seed) damage and the damage to single plants in segregating progenies can be examined, which would not provide enough green bolls for weekly examination.

Bioassay techniques can also be used effectively to this insect. R.L. Wilson and F.D. Wilson (1974) described the use of artificial diets containing lyophilized carpel walls or boll contents. Freshly hatched larvae are placed in diet cups and held until pupation, at which time they are weighed. Pupae are returned to the cups and allowed to develop to adults. Cotton genotypes can be evaluated for effect on larval development time and larval weight, percent pupation, pupal weight, pupation time, and percent adult emergence.

f) Jassids. Jassids (*Empoasca* spp.) inflicts damage on numerous crops. Both field and cage tests are used to screen cotton plants for resistance to jassids. Where natural populations are high, counts of nymphs may be made, either at one time or periodic samplings. Counts are made on a specified number of leaves of randomly selected plants in each plot or entry (Parnell et al. 1949). Jassid nymphs also may be used to infest plants confined in individual cages (Batra and Gupta 1970). The caged insects are permitted to feed, develop into adults, and reproduce for a particular period, after which the individually potted and caged plants are fumigated and the dead insects are counted.

g) Thrips. The germplasm of cassava *(Manihot esculenta)* was screened to thrips (*Frankliniella williamsi* and *Corynothrips* spp.) by Schoonhaven (1974). Part of the collections were evaluated under natural infestation during two successive dry seasons. Plants were evaluated at 4 and at 8 months and the average of these two assessments was used as a resistance classification. Symptoms of damage were classified into six classes

0 = No symptoms.
1 = Yellow irregular leaf spots only.
2 = Leaf spots, light leaf deformation, parts of leaf lobes missing, brown wound tissue in spots on stems and petioles.
3 = Severe leaf deformation and distortion, poorly expanded leaves, internodes stunted and covered with brown wound tissue.
4 = As above, but with growing points dead, sprouting of lateral buds.
5 = Lateral buds also killed. Plants greatly stunted, with witches broom type appearance.

h) Fruitfly. This is one of the destructive pests of cucurbits. Nath et al. (1976a) screened the *Cucurbita maxima* to fruitfly, *Dacus cucurbitae*. In addition, natural populations of fruitfly were placed in each plot to insure infestation and to avoid any escape. The rating was made on the basis of percentage of fruits damaged as follows:

Scale	Category	% Damage fruits/plant
1	Highly resistant	0- 10
2	Resistant	11- 25
3	Susceptible	26- 50
4	Highly susceptible	51-100

Plants were screened three times at intervals of 14 days during the growing period after the appearance of first female flower. Only those fruits in which maggots were observed feeding on the pulp were considered as damaged.

i) Corn Earworm. Newly hatched larvae of corn earworm, *Heliothis zeae*, from a laboratory colony are applied to the silk on each ear of corn with a small brush. This method is time-consuming and laborious. Large populations in a breeding programme cannot be infested artificially. N.W. Widstrom and Burton (1970) used insect eggs for artificial infestation. They reported that two field applications of ten or more eggs at intervals of 0-10 days, or a single application of at least 30 eggs, produced damage to corn comparable to that produced by any method of larval infestation tested. Repeated applications were more important than number of eggs per application in producing results comparable to larval infestation methods.

j) Shootfly. Thus is a pest of the seedling stage and is very destructive in sorghum. Greenhouse techniques require rearing flies, *Atherigona soccata*, on seedling plants in cages using eggs taken from infested plants (Soto 1974). Two-week-old sorghum seedlings grown in flats are placed in adult-infested cages to serve as oviposition sites and hosts for larvae. Flats are removed from the cage and placed outdoors after plants are infested with eggs (about 24 h). When the larvae reach the third instar, infested seedlings are uprooted and placed in flats containing a layer of moist sand. Flats are then held in the emergence units until adults emerge. Seedling resistance is generally based on percent infested seedlings or percent dead hearts. Both resistant and susceptible cultivars are used as check.

k) Podborers. Podborers cause considerable losses to grain legumes. Jackai (1982) developed a method for screening resistance in a large number of collections of cowpea, *Vigna unguiculata* to the legume podborer, *Maruca testulalis* (Geyer). Several damage parameters were measured, including damage to stem, flowers, pods and seeds. Some of these parameters were assessed in more than one way. The stem and pod damage measurements provided the assessment of resistance to the borer at the initial stage. At a later stage, when the number of cultivars has been reduced considerably, larval counts in flowers and seed damage measurements can be included.

l) Stored Grain Pests. A common practice in searching for resistance to stored grain insects is to infest a grain sample of each variety with a specific number of insects for a period of feeding and oviposition. Resistance is evaluated by counting damaged grains and/or first generation progeny. Balancing the relative humidity and seed moisture content of the test samples is necessary in this case.

m) Nematodes. Preliminary screening tests for resistance to potato cyst nematodes (*Globodera* spp.) have been carried out in the glasshouse. These tests consist of growing plants in infested soil in pots and the number of cysts on the outside of each root ball is counted. Those lines/plants which appear resistant in this "root ball" test are then retested more thoroughly, using a soil-flotation method by which cyst numbers can be determined accurately (Howard and Fuller 1975).

Susceptibility of red clover, *Trifolium pratense*, seedling to stem nematode, *Ditylenchus dipsaci*, in the laboratory was tested by Toynbee-Clarke and Bond (1970). The seedlings were grown between layers of moist filter paper and one drop of a suspension of nematode larvae is applied to each seedling; after 2 or 3 weeks, the seedlings are scored for severity of symptoms based on following scoring system:

0 = No swelling.
1 = No swelling but with some necrosis.
2 = Stunted, no swelling.
3 = Slight swelling of hypocotyl.
4 = Greatly swollen petiole.
5 = Greatly swollen hypocotyl.

Seedlings from classes 0 and 1 were selected for further tests and for use in breeding.

Philipps and Dale (1982) described a bulk seedling test, which can be applied shortly after completion of a crossing programme. The test assessed resistance to *G. pallida* in progeny derived from a crossing programme involving *Solanum tuberosum, S. vernei* and *S. tuberosum* spp. *andigena*. It provides good estimates of both the mean level of resistance within progenies and the breeding value of the parents.

In the preceding pages screening techniques for some of the important pests have been dealt with. Testing for resistance to insect pests may be carried out in the field and/or in the glasshouse. In field tests experimental errors are often high because of insufficient number of insects or non-uniform insect damage; however, field trials can be laid out successfully in areas where pest attacks are most likely to occur. This is more successful for soil pests or to air-dispersed pests which occur more frequently in certain localities. It is not possible always to rely on the natural occurrence and the spread of pests. Therefore, infestations under controlled conditions with an equal number of pests are carried out. In plant breeding programme usually large number of germplasm lines are screened. It is desirable first to screen the germplasm in the field under natural infestations and the promising lines may be further tested by laboratory tests.

5.6 Assessment of Insect Pest Resistance

A rapid, repeatable rating scale must be developed that is related to the development of the insect or to the economic damage done by the insect. Sometimes both may be used as criteria in screening plant genotypes. The effect of the plant upon the insect, namely, insect weight, fecundity, growth interval of instars, can be measured, or the effects of the insect upon the plant, such as a damage-rating scale or a recovery rating, can be assayed (J.N. Jenkins 1981).

The pictorial standard of damage may be used at the initial stages to identify rapidly material worth advancing and to differentiate intermediates and susceptibles. It is better to score individual plants rather than a whole plot. The lower scale indicates resistance and the higher indicates susceptibility. However, 0 may be avoided as a score for statistical evaluations. The later evaluation should involve the level and expression of resistance, including the insect counts. It is important that plants of the same growth stage or maturity are evaluated. Seedling stage and adult plant stage resistance may be correlated.

6 Breeding for Resistance to Diseases and Insects

The cultivation of resistant varieties has been recognized as the most effective, ideal and economical method of reducing crop losses (Stakman and Harrar 1957). The breeding for resistance is generally no way different than breeding for other traits. However, in resistance breeding the two biological entities, host plant and parasite, are involved, whereas in breeding for other traits the breeder deals with the variability in test material only. The first step in resistance breeding programmes is the collection of natural variability followed by finding out the sources of resistance. The information on these aspects has been described in detail in Chap. 5. The next step is to incorporate the resistance gene(s) from the donor parent(s) using various methods including induced mutations, where the susceptible alleles are altered by the use of mutagens. These methods are discussed in this chapter. In view of the dynamic nature of parasites, the resistant gene(s) fall susceptible after a few years or are no longer effective. Therefore, the resistance breeding programme is a continuous one. An attempt is also made in this chapter to describe how best the race-specific or non-specific genes may be managed so that the life span of resistant gene(s) is prolonged and also the disease losses are avoided or reduced.

6.1 Methods of Breeding for Resistance

Selection for resistance to pests and diseases is relatively easy, but certain host plant genotypes may show differing reactions to races/biotypes. It is desirable to test host genotypes against a wide range of variants of a parasite before selection is made. This could be done by using variants separately or in known compositions. The use of individual variants is important for studying the genetics of the host-parasite relationship but is not important for practical breeding tests. The mixture of variants can be maintained individually on a range of host genotypes or on culture medium. The breeding material can also be grown at several locations (sites) where different variants of parasites are expected to occur. The coordinated programmes of major food crops in India, conducted by the Indian Council of Agricultural Research, is one of the best examples of testing crop varieties under different agro-climatic conditions.

Although similar procedures are adapted in selecting for pest and disease resistance in self- and cross-pollinated crops, there are important differences in handling the two types of crop. In self-pollinated crops individual plants are selected to form a variety, while in cross-pollinated crops this is seldom a basis for selecting resistance. This is because of self-incompatibility, inbreeding depression etc. The flower structure of self-pollinated crops poses problems in intermating, and consequently the breeding methods like recurrent selection are of limited use; however, with the use of genetic male sterility they may be used in autogamous crops too.

6.1.1 Cross-Pollinated Crops

Several methods can be used in allogamous crops for improving populations for developing varieties with resistance to pests and diseases. It appears from the literature that mass selection/recurrent selection has been used widely. Therefore, it will be discussed in detail and other methods will be dealt in brief.

a) Mass Selection/Recurrent Selection. This is one of the most commonly used breeding methods. It consists of selecting individual plants for resistance from the heterozygous population of plants which is inoculated/infested artificially. The population from the selected plants is reinoculated in the next generation, and susceptible plants are eliminated before intermating to produce seed. Since resistance to pests and diseases has high heritability, mass selection would be useful as a breeding tool. Mass selection has been used to develop resistant varieties for anthracnose and wilt disease in alfalfa (Allard 1960). The object is to achieve a higher proportion of resistant plants in each successive generation (i.e. by recurrent selection).

Recurrent selection has been used most effectively to improve resistance to pests and diseases in forage crops and in other cross-pollinated crop species. M.T. Jenkins (1940) outlined the procedure of this method and M.T. Jenkins et al. (1954) used it for concentrating genes for resistance to *Drechslera turcica*, the leaf blight disease in corn. Three generations of selection were found to be effective. Penny et al. (1967) used recurrent selection successfully in corn for shifting frequencies of resistant genes to a higher level for resistance to the first brood of the European cornborer *(Ostrinia nubilalis)*. Two cycles of selection were found adequate and three cylces of selection produced essentially borer-resistant cultivars. A recurrent selection breeding procedure was used to reduce leaf-feeding damage by first-generation European cornborer and to increase DIMBOA content in a synthetic maize cultivar (BS_1) (Tseng et al. 1984). Gene action for these traits was found to be additive. Recurrent selection was used to increase earworm, *Heloithis zeae* resistance in corn (N.W. Widstrom et al. 1970) and at the same time desirable agronomic traits such as yield were maintained. Osuna et al. (1983) reported its use for earworm resistance in two corn populations Dent and Flint composites, grown in isolation for two cycles. Planting dates were adjusted to enhance the chances for high infestations of earworm. From each cycle 200 S_1 progenies were selected for ear damage, husk diameter and ear weight. Recurrent selection in both composites was effective. For Dent composite, the percentage of ears with kernel damage was reduced from each 5.56 and 5.72% as observed and estimated, respectively. The percentage of ears with earworm damage for flint composite, the average observed and estimated reduction for each cycle was 6.93 and 7.01%, respectively.

Recurrent phenotypic selection has been an effective method for increasing disease and insect resistance in alfalfa, *Medicago sativa*. The published information on this form of mass selection in alfalfa is obtained from two broad-based populations, i.e. A and B germplasm pools. The effects of seven cycles of selection in these two pools for resistance to potato leafhopper yellowing caused by *Empoasca fabae* and to leaf rust, *Uromyces striatus*, were described by Dudley et al. (1963). The results have been described for subsequent cycles of selection for resistance to potato leafhopper yellowing by Hanson et al. (1963); spotted alfalfa aphid *(Therioaphis maculata)*; and common leaf spot *(Pseudopeziza medicaginis)* were described by J.H. Graham et al. (1965). Hanson et al. (1972) used it for the development of alfalfa populations resistant to four diseases, namely, rust, common leaf spot, bacterial wilt and anthracnose; and to two insect pests, namely, spotted alfalfa aphid and potato leafhopper. There was an increase in the general vigour of populations simultaneously in the generation of selections conducted in the field. Selection was highly effective for increasing resistance to each disease or insect. The selection for resistance to a particular pest did not cause detectable reductions in levels of resistance to other pests developed in the previous cycles (R.R. Hill et al. 1969).

Three cycles of recurrent selection for anthracnose *(Colletotrichum trifolii)* resistance in three alfalfa populations increased the level of resistance from 4.2 in the original to 1.7 after selection (1 = highly resistant and 5 = dead plants). More importantly, the stands for the selected populations were 85% to 95% alfalfa, whereas in stands of the original populations, 50% were alfalfa plants and 50% were weeds (Devine and Hanson 1973).

D.K. Barnes et al. (1971) used A-C4 and B-C 4, the two unrelated populations of alfalfa, and subjected then to recurrent cycles of phenotypic selection for resistance to bacterial wilt caused by *Corynebacterium insidiosum*. Thirty five hundred to 5000 seedlings were inoculated in each populations per cycle. About 150 plants in each population were intercrossed to initiate each new cycle of selection. The disease severity indices (0-5 scale, where 0 = healthy, and 5 = dead) for A-C4 and after one and two cycles of selection were 3.72, 2.53 and 1.38, respectively, for B-C4 and after each of the four cycles of selection, 4.25, 3.94, 3.67, 3.13 and 2.63, respectively. Recurrent selection was effective for developing resistance to bacterial wilt.

The phenotypic recurrent selection in alfalfa (Hanson et al. 1972) for resistance led to the release of alfalfa cultivars Williams, Cherokee, Team, and Arc that were derived directly from recurrent selection.

Recurrent selection in the progeny of crosses involving Purple Top White Globe, Raab Salad, and Shogoin turnip (*Brassica campestris* Rapifera group) resulted in considerable progress towards developing cold-tolerant and turnip aphid *Hyadaphis erysimi* resistant turnips (W.C. Barnes and Cuthbert 1975). The aphid resistance seemed to be based on non-preference and possibly antibiosis and tolerance.

The use of recurrent selection to concentrate polygenic resistance for intra-population improvement was used in pearl millet, *Pennisetum typhoideum* (Purm S & H) against ergot, *Claviceps fusiformis* (Loveless) resistance. Three and four cycles of selection had increased the proportion of plants with 0-5% ergot severity (Chahal et al. 1981).

Estimates of among and within half-sib family variance components for disease score in each of the two corn *(Zea mays)* populations were obtained from artificially inoculated experiments in 1, 2 or 3 years. These estimates were used to predict the response of mass, half-sib and S_1 selection for resistance to northern corn leaf blight

(Drechslera turcica); Diplodia stalk rot *(Diplodia maydis)*, anthracnose leaf blight and stalk rot *(Colletotrichum graminicola)*. In addition, among half-sib family variance component estimates for yield were obtained from uninoculated experiments for the same families. Genetic correlations among yield and reactions to the four diseases were estimated. Genetic correlations between yield in the absence of disease and disease reactions were all near zero and non-significant. All significant correlations among disease reactions were positive. These results suggest that mass selection would be the most efficient method of improving disease resistance in corn populations and would not affect yield if adequate population size is maintained (Miles et al. 1980).

On the basis selection advance per year, the most efficient procedure involved randomly intermating a group of potato clones, planting seedling tubers in the field test to measure potato leafhopper, *Empoasca fabae* resistance to infestation, and then bulking the seed from the most resistant clones to complete the selection cycle in 1 year (Sanford and Ladd 1983). A population developed by five cycles of selection for resistance to infestation using this procedure resulted in a decrease of 57% in the level of infestation from the original population.

b) Line Breeding. Selected plants are either selfed or inter-pollinated, and the resulting progenies or lines are individually tested for resistance; only the most resistant lines are retained for subsequent breeding. These are then inter-pollinated to produce a composite cross.

c) Polycross. This method consists of selecting resistant plants from a heterozygous population and inter-crossing these or inbred lines derived from these, in all possible combinations. The progenies of a polycross can be bulked. Resistant plants are subsequently selected from the bulk population, or the progenies of individual lines can be tested separately.

d) Synthetic/Hybrid Varieties. Resistant lines resulting from line breeding or recurrent selection programmes can be used to produce hybrid or synthetic varieties. A synthetic variety is produced by intercrossing a number of selected plants, lines or clones which have been found to be good combiners. Hybrid varieties are produced by controlled pollination between lines. The parental lines ought to be maintained separately so that synthetic or hybrid varieties may be reconstituted as required. Inbred lines may be improved for resistance by using the back-cross method or convergent improvement. This may be done by using male sterile cytoplasm.

Rowe and Hill (1981) compared inter-population improvement methods for alfalfa. Top-cross selection was more effective than polycross selection for resistance to *Colletotrichum trifolii*, when the tester parent had less resistance than the base population, and top-cross selection was less effective when the tester parent was more resistant.

6.1.2 Self-Pollinated Crops

In self-pollinated crops pedigree and backcross methods with certain modifications as per convenience of the breeders are used frequently and other methods are not frequently used. Therefore pedigree and backcross methods are described in detail along with examples and other methods are discussed in brief.

a) Mass Selection. Plants of a similar phenotype are selected for resistance from a population and the progenies are bulked to form the basis of variety. Such varieties are easy to produce. Their heterozygosity might give them certain advantages over pure line varieties for pest and/or disease resistance.

b) Pure Line Selection. These are each derived from the progeny of a selfed homozygous plant selected from a land variety or commercial variety. The progeny is retested for resistance and evaluated for other desirable characters in succeeding generations and, if promising, it is multiplied to produce a new variety. This is a simple and quick procedure, and many useful varieties of self-pollinated crops have been developed from "off-types" which presumably arise either by mutations or by natural hybridization. In the 1930, F.A. Wagner of the Kansas Agricultural Experiment Station, USA found two plants of Dwarf Yellow Milo (sorghum) variety that were apparently normal in every way, but could be grown of infested soil with root rot or milo disease caused by *Periconia circinata*. A resistant strain from these plants was developed (Allard 1960).

c) Hybridization. This involves crossing of two pure line varieties with the objectives of transferring resistance from donor parents and of combining characteristics from each parent. It enables the breeder to combine resistance to several races or biotypes or to different pests for diseases in a single variety. The F_1 plants from a cross are identical in their genetic constitution. Segregation occurs in the F_2 generation onwards and homozygosity is regained in succeeding selfed generations. The transregressive segregants for resistance may occur with quantitatively inherited resistance, such segregants may be selected in the F_2 and later generations. Selection after hybridization is based on either of the following methods.

Pedigree Method. If a resistant parent can contribute to improved adaptation, yield, yield components and quality etc., the pedigree method of handling the segregating generations may be followed. It consists of selecting individual F_2 plants for desirable features, including resistance to pests and diseases. The progenies of these selections are reselected in each succeeding generation until homozygosity is obtained. Artificial epiphytotic conditions are created for pest(s) or disease(s) in the early segregating generations (as described in screening techniques).

The pedigree method is used in handling the early generation breeding lines in rice (Khush 1971). A portion of seeds from plant selections is used for planting the pedigree rows and the rest for screening of disease and insect resistance.

Dwarf plants from the cross of susceptible (dwarf) and resistant (tall) varieties of rice were selected in the F_3 generation. The progenies of these plants were screened up to F_3 generation for pest and disease resistance. Several selections at IRRI were observed to be resistant to common pests and diseases (Pathak 1971). Dwarf lines with improved plant type are screened for resistance traits of the tall parents in these crosses (Khush and Beachell 1972).

The pedigree method has been utilized in screening sorghum lines against shoot flies (Blum 1972). During F_3 and subsequent generations, promising resistant lines were test-crossed on male sterile CK 60 to identify restorer genes and to evaluate combining ability for yield. Some resistant lines (F_5) giving non-restorer reaction were back-crossed to male sterile CK60 in order to develop resistant male steriles. Some of the final F_7 resistant lines effectively maintained non-preference and tiller survival resistance.

The pedigree method has been very widely used in breeding for disease and insect resistance and the majority of resistant varieties of self-pollinated crops have been developed by this procedure.

Bulk Population Method. This method is also suitable for combining characteristics from both the parents. In this method the early segregating generations (F_2-F_6) are bulked together without selection. In later generations when most plants are homozygous, individual plants are selected for resistance and their progenies are evaluated as in the pedigree method. Artificial epiphytotic conditions are created in the later generations for selecting resistant plants/progenies.

Mass Pedigree Method. Some aspects of the pedigree and bulk population methods are combined in this method. Hybrid populations are grown in bulk until a pest or disease attack occurs naturally, when single plants are selected for resitance. At one extreme, bulking may end in the F_2 generation, hence it is no way different from the pedigree method, and if bulking may be continued for many generations in the absence of suitable conditions for selection, it is similar to the bulk population method. The individual plants selected in either case are handled as pedigree method.

Back-Cross Method. If the resistant parent is a wholly unadapted type, the back-cross method is appropriate to transfer certain characteristic of a line or variety into breeding material. It is more useful to transfer one gene or a few genes, such as major genes conferring a high degree of resistance, from one genetic background to another. The genes from wild species have often been transferred to susceptible but otherwise satisfactory varieties by this method. Where resistance is dominant, the F_1 is back-crossed to susceptible parent. The progeny of the first back-cross generation is tested for resistance. The resistant plants are back-crossed to the susceptible parent (recurrent parent). After several generations of back-crossing (generally five to six), plants with characters almost identical to the original susceptible variety are obtained. These plants have the added advantage of resistance genes. Where reistance is recessive, the progeny of each back-cross generation is selfed to obtain homozygous recessive plants. This method is not suitable for quantitatively inherited resistance, which is controlled by polygenes.

The wheat varieties Poso 42 and Big Club 43 resistant to hessian fly were developed by the back-cross method and replaced the susceptible varieties in California. Dual Wheat, resistant to this pest, was released in Indiana (R.M. Caldwell et al. 1959).

The back-cross method can also be used for transferring resistance to more parasites. Once a variety has been improved for resistance to one parasite, it can serve as the recurrent parent for incorporating resistance to the second one.

The procedure adapted for the development of the Baart 38 variety of wheat, with resistance to stem rust and bunt diseases, is described by Allard (1960) in detail. The stem rust resistance was transferred from Hope wheat to Baart and merging this type with bunt-resistant Baart to form Baart 38. Several examples of using the back-cross method are discussed by F.N. Briggs and Allard (1953).

Sometimes it is difficult to recover lines that have a good plant type from single crosses following the pedigree method. One or two back-crosses to the improved plant-type parent are, therefore, imperative. Occasionally an F_1 is crossed to a third improved plant-type parent to obtain a three-way cross (Khush and Beachell 1972). The back-cross method is also useful for developing isogenic or near-isogenic lines of the multiline cultivars.

d) Mutation Breeding. The resistance gene(s) sometimes are not available in cultivated types and the wild germplasm is not easily crossable. Often the cultivated variety is higher-yielding and widely adapted, but is susceptible to a particular parasite or a race/biotype and the breeder is interested to alter the susceptible alleles to resistance without changing its other traits. Under these circumstances mutation breeding for resistance is the obvious choice.

W.A. Orton (1900) selected surviving cotton plants on soil heavily infected with cotton wilt, *Fusarium oxysporum*. Bolley (1905) also obtained flax resistant to *F. oxysporum* f.sp. *lini* under similar conditions. Resistance may sometimes first be detected under conditions when a disease is so severe that the crop is almost completely destroyed (P.R. Day 1974). Such resistant plants could be due to spontaneous mutations; however, chance out-crossing or seed admixture cannot be ruled out. The mutations can also be induced artificially with the use of mutagens. Konzak (1956) presented evidence that disease resistance in higher plants is obtainable via induced mutations. The mutagens could be physical or chemical.

Micke (1983) observed that if a mutation-treated population is subjected to infection by a particular pathotype, it is possible to find resistance differences only with regard to that pathotype. However, if a mixture of pathotypes is used, it may be able to detect resistances against all components of this mixture. If disease/insect pressure is very high from using artificial screening techniques, it will not be possible to detect mild/non-specific forms of resistance, which may be of much practical value. Therefore, he suggested that natural infection pressure may be more appropriate.

A.T. Wallace and Luke (1961) reported that the frequency of obtaining a resistance gene by a mutagenic treatment is extremely low (1 to 35×10^{-8}). It may therefore be useful to observe a large number of plants under natural infection, when manpower is limited. Murray (1969) recovered 12 mutants with moderate resistance to *Verticillium* in peppermint, *Mentha piperita*. Approximately 100,000 plants derived from neutron or X-ray-treated stolons, were grown in wilt-infected soil. The mutants were grown in wilt-infested soil. The mutants were presumed to be dominant. An induced mutant, called m1-0 (Favret 1971a), for resistance to *Erysiphe graminis* f.sp. *hordei* in barley occurs with an unusually high frequency. The mutant is recessive and has the pleiotropic effect of producing necrotic flecking in the absence of the pathogen and is associated with a small reduction in yield. Jørgensen (1971) noted that plants carrying m 1-0 were not susceptible to any physiological race of mildew in large-scale tests in several parts of the world.

The general outline of a mutation breeding procedure in self-pollinated crops is presented in Fig. 6.1. The same procedure may be followed in cross-pollinated and vegetatively propagated crops. In cross-pollinated crops, instead of selfing individual plants in M_1 generation, a group of plants are selfed together. In asexually propagated crops cuttings are used in place of seeds.

Exposure/treatment of seeds with physical/chemical treatment.

Each M_1 plant is selfed. Screening of dominant mutations, if any may be carried out. Seed from individual plant/spike/tiller may be harvested separately.

M_2 plant to progenies are grown. Selection of individual plants with moderate to high degree of resistance is carried under epiphytotic conditions (Fig. 6.2).

M_3 plant to progenies of selected M_2 plants are grown. Selection of plants/progenies under epiphytotic conditions.

M_4 generation consists of progenies of individual selected M_3 plants and selected progenies. Preliminary yield trial of selected progenies along with parental variety and standard checks are conducted.

Multi-location trial of the high yielding and disease resistant progenies alongwith checks.

↓

Release and seed multiplication of superior progeny.

Fig. 6.1. General outline of a mutation breeding procedure in autogamous crop plants

Sigurbjornsson and Micke (1969) listed 77 cultivars that were developed as a direct consequence of induced mutations; 16 of these were cited as having improved disease resistance. Recently Micke (1983) reported that 225 varieties of agricultural and horticultural crop species have been derived from mutation induction. At least 58 of them are with improved disease resistance (14 varieties of barley, eight of bean, six of wheat, six of oats, four of rice, four of durum wheat, as well as varieties of soybean, jute, mustard, cotton, peppermint, millet, sugarcane, lespedeza and apple). Regarding insect resistance, the known mutant rice variety is Atomita 2, released in Indonesia, which is supposed to have resistance against the brown planthopper (Micke 1984, personal communication).

6.1.3 Vegetatively Propagated Crops

Some of the important crops like sugarcane, potatoes, strawberries, raspberries, pineapples, bananas and most of the fruit trees are propagated through vegetative or asexual organs. The plants of these crop species are usually heterozygous. The vegetative propagation maintains this heterozygosity. If the desired level of resistance to parasite is obtained, it can be maintained easily. Resistance is sought either by testing mixed populations of clones from different sources of resistance or by selecting the best progenies obtained by hybridizing different clones. Segregation starts in the F_1 itself, therefore each F_1 plant must be tested separately for resistance.

Grafting has been used as a method of controlling insects the grape aphid, *Phylloxera vitifoliae*.

Fig. 6.2. A desirable mutant soybean plant with resistance to MYMV, selected in the gamma-ray irradiated M₂ population of greengram

6.2 Management of Disease and Insect Resistance

The mode of inheritance of resistance in the host may be race-specific (vertical) or non-specific (horizontal). Depending on the mode of inheritance, several methods have been proposed for the better utilization of the resistance gene(s). Therefore, in the ensuing pages how best the vertical and horizontal types of resistance genes could be manipulated for the control of diseases and insect pests are discussed. These techniques can prolong the average life span of resistant gene(s) and at the same time reduce the risk of catastrophic losses.

6.2.1 Management of Vertical Resistance Gene(s)

a) Recycling and Sequential Release of Resistance Gene(s). This is based on the same principle as the common practice of crop rotations to control certain soil-borne or root-infecting pathogens. Stevens (1949) suggested that a system of variety rotations, similar to crop rotation, should be followed so that inoculum of a particular race(s) does not build up in sufficient quantities over a period of time to create an epidemic. The varieties rotated should, however, have different genes for resistance. According to Stevens, where varieties are replaced more frequently, with each rapid increase in the numbers of the "new" races due to new host varieties, there is corresponding decrease of "old" races. After 5 or 10 years of widespread planting of a new host variety, it may be that the formerly well-known races or certain diseases will have become scarce, and that the older host varieties can be replanted with profit.

Rao (1968) noted that the system proposed by Stevens may be worth consideration only when: (1) the old varieties are resistant to new races and new diseases that have become important, (2) the populations of the old races are not high at the time of replanting old varieties, (3) the new varieties are not superior to the old varieties in yield, quality and other agronomic attributes; and (4) seed multiplication and distribution is not cumbersome. Since in an on-going breeding programme improvements in cultivars are always being made, the utility of the system proposed by Stevens may be limited. However, if a judicious back-cross programme is followed keeping the original genotype, with additional genes for resistance added, as new races came up, the system could be balanced.

Release one gene for resistance and wait until it becomes ineffective; release the second gene and so on. This approach was adopted to control stem rust of wheat in Australia between 1938 and 1950 (I.A. Watson and Luig 1963). The sequential release strategy has been employed for resistance to brown planthopper *(Nilaparvata lugens)* at IRRI (Khush 1979). IR 26 and IR 1561-228-3 varieties of rice with resistant gene Bph 1 were released in 1973 and 1974 respectively. Towards the end of 1975 and in 1976, these varieties started to show susceptibility at some locations in the Philippines. But by that time multiple disease- and insect-resistant varieties with bph2 resistance gene for BPH had become available (IR36 and IR38) and were released as replacements for the varieties with Bph 1. Later on at IRRI, breeding lines with Bph3 and bph4 for resistance to BPH were available. A proposal for race prediction and gene rotation to keep development of new disease-resistant varieties of rice for farmers ahead of the development of new races of disease organisms was advocated at IRRI (1980) (Fig. 6.3). The system consists of the following steps:

1. Identify the monogene for resistance that is effective against races of the pathogen present in a specific cropping area.
2. Use of monogene (identified in step 1) for resistance to develop varieties for the specific cropping area and release these varieties. This will reduce and possibly eliminate the presently occurring races in the cropping area.
3. Test the resistant monogene at sites remote from the specific cropping area, through international testing programme. The race of pathogen that overcomes the monogene in one or more of the remote areas will be representative of the race that will

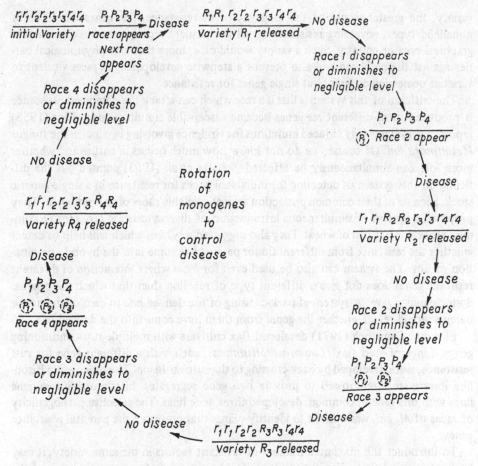

Fig. 6.3. Proposal for race prediction and gene rotation to keep development of new disease resistant varieties for farmers ahead of the development of new races of disease organism (adapted from IRRI, 1980). P = specific gene for pathogenicity in the pathogen population; p = allele of P incapable of being pathogenic; R = specific gene for resistance in a rice variety; r = allele of R resulting in susceptibility; $\widehat{P}$ = broken circle designates presumed fate of P gene in the pathogen population as disappearing or becoming much reduced in frequency

eventually evolve in the cropping area following the introduction of the new varieties containing the monogene (step 2).

4. Identify, based upon results obtained from the programme (step 3), a monogene for resistance to the race predicted to occur in the cropping area and use it in the variety development programme.

b) Pyramiding of Resistance Gene(s). Simultaneous introduction of diverse genes for resistance into the cultivar was proposed for the first time by I.W. Watson and Singh (1952). Athwal (1953) gave a scheme for introducing two genes, following a system of back-crossing and taking advantage of epistatic reaction types. The system is based on the mutation in the pathogen at more than one locus being much rarer than the mutation at one locus. The higher the number of diverse genes introduced into a single variety, the greater would be the longevity of its resistance. The genes should be of nonallelic types, governing resistance to all the virulent races present in a given geographical area or country. Such a variety would offer more than one physiological barrier against the pathogen and also prevent a stepwise development of races virulent to varieties possessing different but single genes for resistance.

The criticism of this system is that if a race which can attack the combined resistance is produced, then different res genes became susceptible simultaneously. Flor (1958) reported simultaneously indaced mutations for virulence involving two loci in the fungus *Melampsora lini*. Of course, we do not know how much occurs in nature and whether more loci can simultaneously be affected. Schafer et al. (1963) pointed out the difficulty in this system of detecting the individual genes for resistance in a single genetic stock, because of their common protection against available races of the pathogen. They proposed a system of simultaneous introduction of diverse resistant genes into a common stock for leaf rust of wheat. They also suggested a system which will help to detect whether the resistance from different donor parents has come into the hybrid combination or not. The system can also be used even for cases where interaction of different resistant genes does not give a different type of reaction than that which the original donor parents gave. A system of back-crossing of the derived line to each of the donor parents would reveal whether the genes from them have come into the derived line.

Flor and Comstock (1971) developed flax cultivars with multiple rust-conditioning genes. Lines of seed flax *(Linum usitatissimum)*, each with a different gene for rust resistance, were developed by back-crossing to the cultivar Bison. The monogenic Bison-like lines were inter-crossed to provide two gene segregates. Inter-crossing two gene lines with one gene in common, developed three gene lines. The selective pathogenicity of races of *M. lini*, was utilized to identify plants that possessed the parental resistance genes.

To introduce the maximum number of resistant factors in the same variety, it may be kept in mind that when the virulence factors necessary for giving evidence of the corresponding resistance genes are not available, it is rather easy to lose some of the latter (resistance genes) (Zitelli and Vallega 1977), for instance line Frontana, K.58-NT II 50-35 of wheat has resistance in many parts of the world and seems to have no less than five resistance factors (Sr 5, Sr 6, Sr 8, Sr 9, S 11) effective with respect to the races of *Puccinia graminis* f.sp. *tritici* present in Italy (Zitelli and Vallega 1971). Through back-crossing the authors have tried to transfer all these factors to some Italian bread wheat varieties. Genetical analysis made on the offspring has shown that not all of these factots carried by parental variety had been inherited. This was because they did not have the indispensable virulent factors at their disposal.

Similar erosion during the selection process has happened with the durum wheat Yuma (Zitelli and Vallega 1968) which has at least four resistance factors with respect to stem rust. Some of the lines derived from crosses with susceptible varieties have inherited all of these factors, while others are susceptible to some races. The resistance of HD 2009 to black rust was different from that of Tenzano Pintos Precoz (TzPP). Its recessive factor was common with HD 2189 and complementary factors common with MJ1 and E 8643. Resistance of HD 2177 was not common with the three varieties. Similarly, resistance in HD 2189 showed diversity from the three varieties. Brown rust resistance of all three varieties HD 2009, HD 2177 and HD 2189 was due to factors different than those in TzPP MJ1 and E 8643. Because of such diversity of genes it should be possible to pyramid and combine genes into new cultivars of wheat to provide sustained resistance (Kaushal and Upadhyaya 1983).

At IRRI, pyramiding of two major genes for BPH resistance is also being tried (Khush 1979). Bph 1 and bph 2 are closely linked (Athwal et al. 1971). Similarly, Bph3 and bph4 are also linked (Lakshminarayana and Khush 1977). However, bph 1 and Bph 3, Bph 1 and bph4, bph2 and Bph3 and bph2 and bph4 segregate independently of each other and can be combined (Khush 1977b). The rice varieties with two genes for resistance to BPH will have a longer useful life (Khush 1979).

c) **Regional Deployment of Resistance Genes.** Resistant varieties with different resistance genes should be developed and recommended for different geographical regions of the country, where the crop covers a sizable area. As pointed out by R.R. Nelson (1972), this type of gene deployment is essentially a geographical multiline. A formal plan for regional deployment of genes is in effect for resistance to crown rust fungus in oats in Iowa (Frey et al. 1973).

M.S.S. Reddy and M.V. Rao (1979) suggested a strategy for controlling leaf rust, *Puccinia recondita* f.sp. *tritici*, of wheat in India by the use of regional deployment of resistance genes. India could be divided into three regions, (A) central plains, (B) northern Himalayan, and (C) southern Nilgiri and Palani Hills. Stronger genes (Lr 9 or Lr 19 or Lr 10 + Lr 15) can be deployed in region A, Lr 1 + Lr 10 + Lr 1b or Lr 3 ka + Lr 10 + Lr 17 in region B, and Lr 3 ka + Lr 10 + Lr 20 or Lr 1 + Lr 10 + Lr 17 in region C. When a number of genes are in operation, the possibility of build-up of a super-race is minimized, and the pathogen will became stabilized. They suggested that a similar line of work with Sr and Yr isogenic lines will help in the release of stable varieties for all the three prevalent wheat rusts in India.

This approach can be followed for any other crop diseases and insects when enough genes are identified.

d) **Chromosome or Genome Substitutions.** For breeding resistant varieties, the sources of disease resistance are sought and the resistant gene(s) are incorporated in cultivated varieties. If gene(s) for resistance are not available in the cultivated species, the breeder transfers resistance from related species/genera. Due to genetic or cytoplasmic or genome cytoplasm interactions, the chromosomes of different species fail to pair either completely or partially, resulting in sterility of F_1. To facilitate recombination, either the F_1 or the amphidiploid is back-crossed with the recurrent parent. In this way the whole genome or a whole chromosome or a segment of a chromosome of a recurrent parent is replaced by the donor parent. These techniques, along with appropriate example(s), are discussed in the following sub-sections.

The genome of a wild or related species (donor parent) could be incorporated into the susceptible cultivar of a cultivated species (recipient) through hybridization and chromosome doubling, using colchicine. In general, the species having higher ploidy level is used as female to avoid nucleo-cytoplasmic interference. By using colchicine, amphidiploids are produced which have the full normal diploid genome of both the species. Resistance to clubroot *(Plasmodiophora brassicae)* has been transferred from the turnip *(Brassica campestris)* (2n = 20, AA) to the Swede turnip *(B. napus)* (2n =

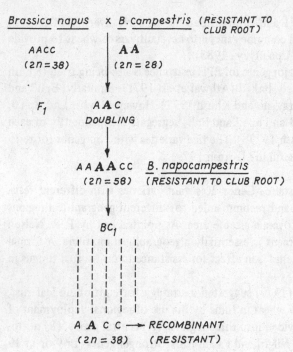

Brassica napus x B. campestris (RESISTANT TO
                                   CLUB ROOT)

Fig. 6.4. The substitution of a whole
genome for developing club root resistant varieties in turnip

36, AACC). Johnston (1974) used colchicine to double the F_1 hybrid to produce an amphidiploid which can be maintained indefinitely and is more easily back-crossed to *B. napus*. Resistant second back-cross progenies were virtually identical with the recurrent parent (Fig. 6.4).

The substitution of a whole chromosome pair of the recurrent parent (susceptible) by one from the donor parent has been used. The wheat streak mosaic disease is a threat to production of winter wheat in many areas of North America. The virus is transmitted by the wheat curl mite *(Aceria teelipae)*. Resistance to virus has not been found in *Triticum* species, but immunity and resistance have been found in species of *Agropyron*. Larson and Atkinson (1973) reported a *Triticum* X *Agropyron* line (2n = 42) in which chromosomes of 4D, 5D and 6D of wheat were substituted by homologous chromosomes from *A. elongatum* (2n = 70, EE), which was found to be immune to wheat streak mosaic. They studied the role of different chromosomes in wheat streak mosaic resistance. They developed different substitution lines for 4D, 5D and 6D chromosomes of wheat. The substitution of wheat by a chromosome of *Agropyron* has nothing to do with resistance to wheat streak mosaic. Substitution of 5D chromosome delays the development of disease, but substitution of 6D chromosome introduces a considerable resistance. The procedure for the substitution of a chromosome pair of wheat by rye is presented in Fig. 6.5. The plants with AA BB DD $R_1 R_1$ can be crossed with different monosomics of cultivated wheat. The monosomic line, where the loss of a full chromosome pair of cultivated species and addition of a pair from other species do not have any adverse affect on agronomic characters, is selected. The addition of the alien chromosome may carry some undesirable characters in addition to resistance gene.

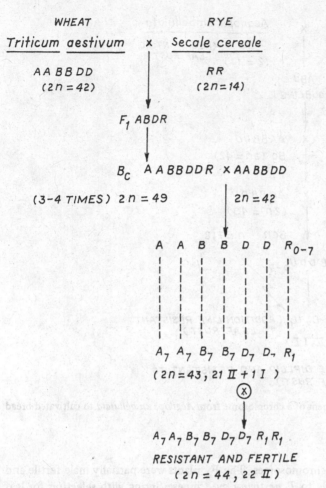

Fig. 6.5. The substitution of a chromosome pair for incorporating resistance from rye to wheat

The transfer of a segment of a chromosome carrying the resistant gene(s) from wild species to a cultivated one is used for developing resistant cultivars. By this method, the single chromosome of the wild species is incorporated into the cultivated species, as in the substitution of the whole chromosome. The trisomic plants are treated with X-rays or another source of irradiation prior to meiosis. These plants are used as pollen parent, so that gametes carrying the haploid complement or the gametes which carry small translocations are more competitive than the gametes carrying the whole extra chromosome. By this method higher frequency of translocations for resistance could be obtained. Such translocations could be terminal, involving small segment or full arm or intercalary one.

Sears (1956) developed the first procedure that used radiation to transfer disease resistance from a chromosome of one species (*Aegilops umbellulata*, $2n = 14$) to a chromosome of another. *A. umbellulata* has good resistance to leaf rust, but the cross between it and *Triticum aestivum* does not produce viable seeds. He therefore crossed *A. umbellulata* with *T. dicoccoides* ($2n = 28$) and produced amphidiploid which could be crossed with *T. aestivum*. In the latter hybrid, the *A. umbellulata* chromosome did

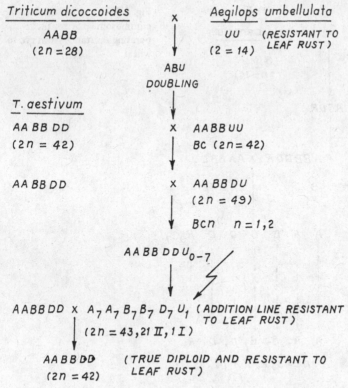

Fig. 6.6. The transfer of a segment of a chromosome from *Aegilops umbellulata* to cultivated bread wheat

not pair with *T. aestivum* chromosomes. The F_1 plants were partially male fertile and two back-crosses were made to *T. aestivum* cv. Chinese Spring with selection for leaf rust resistance. A resistant plant having an added *A. umbellulata* isochromosome was obtained. Its progeny were irradiated with X-rays prior to meiosis and then used as pollen parents in cross with Chinese Spring (Fig. 6.6). Pollen carrying the *A. umbellulata* isochromosome rarely functioned and was therefore screened out. However, wheat-*umbellulata* translocations were often less deleterious, and many of the rust-resistant progeny carried translocations. Only one of the translocations proved to be transmitted entirely normally through the gametes. In this line, later named Transfer, the transloca-tion involved wheat chromosome 6B (Sears 1966). Athwal and Kimber (1972) showed that the *A. umbellulata* chromosome is homeologous with group 6 chromosomes in wheat. Knott (1971a) made nine back-crosses to add leaf rust resistance of Transfer to the cultivar Thatcher. Although the back-cross lines had 1% higher protein content than Thatcher, they were very inferior in baking quality.

Knott (1961) used the procedure of Sears to transfer stem rust resistance from *Agropyron elongatum* ($2n=70$) to a wheat *(T. aestivum)* chromosome. In this case there is no problem in making the cross, but *Agropyron* chromosome did not pair with wheat chromosome in either addition or substitution lines. One of the translocations was transmitted normally through the gametes and has been used as source of stem rust resistance in the Australian cultivars Eagle and Kite. Knott (1964) and R. Johnson

(1966) showed that the *Agropyron* chromosome was homeologous with wheat group 6 and the translocation that was transmitted normally through the gametes was proved to involve chromosome 6 A.

A chromosome carrying leaf, stem and stripe rust *(Puccinia striiformis)* resistance in *Agropyron intermedium* was transferred to wheat by Wienhues (1966). Radiation was used to translocate the leaf rust resistance to a wheat chromosome (Wienhues 1967), but in general the lines carrying translocations gave reduced yield (Wienhues 1973).

Knott (1961), D. Sharma and Knott (1966) and Wienhues (1966) used essentially the same basic procedure as that of Sears (1966), except that they used different sources of resistance.

To prove that irradiations cause translocations rather than mutation: (1) in every case the type of reaction to race 15B in the translocated lines was identical to the lines carrying the full *Aegilops* or *Agropyron* chromosome, (2) the frequency of resistance line was greater than would be expected by mutations, and (3) reduced transmission through the gametes is more likely to occur with a translocation involving one addition of *Agropyron* chromatin and deletion of wheat chromatin.

Ideally, the alien chromosome segment should be as small as possible, particularly if the two species show extensive chromosomal differentiation. Otherwise, the substitution of an alien chromosome segment for a chromosome segment of the recipient species may result in undesirable duplications, deletions and linkages of genes (Knott and Dvorak 1976). The effect is large in diploids, but is often reduced in polyploids, where the effect of single chromosmes is smaller because of the duplication of genetic material in two or more different genomes. The removal of deleterious genes linked to a desired gene is possible if the chromosomes involved are homologous and crossing-over occurs. However, it becomes increasingly difficult as the degree of differentiation increases and the chromosomes involved are only homeologous and not homologous.

e) Multi-Line Cultivars. A multiline cultivar is a population of plants that is agronomically uniform but heterogeneous for genes that condition reaction to a disease organism (Frey and Browning 1971). It has long been recognized that crop homogeneity is a condition favouring epiphytotics (Roane 1973). The use of multiline varieties in self-pollinating crops has been advocated since the late nineteenth century (Browning and Frey 1969, Simmonds 1962). The programmes of multi-line production are based on two radically different philosophies for disease control (Marshall 1977). In one approach, designated as the "clean crop" approach, all component lines of the mixture would be resistant to all prevalent races of the disease (s) to be controlled (Jensen 1952, Borlaug 1959). The aim of this scheme is to keep the crop as free of disease as possible, and at the same time to reduce the threat of catastrophic disease losses following shifts in the racial composition of the pathogen population. In the second approach, designated the "dirty crop" approach (Marshall and Pryor 1979), each line in the mixture also carries a different single gene resistance; however, none of the lines is resistant to all known races of the pathogen. Frey et al. (1973, 1975) have argued that such multi-lines should protect the crop in two ways:

– They should stabilize the race structure of the pathogen population (Suneson 1960, Jensen and Kent 1963, Leonard 1969a,b). This is based on the fact that stabilizing selection against races carrying multiple genes for virulence (Van der Plank 1963,

1968) will ensure that simple races, carrying a single virulence gene, dominate the pathogen population.
- Since each component of the multi-line would be attacked by only one race of the stabilized pathogen population, the remaining lines would act as spore traps, reducing the rate of spread of the disease. In this way, multi-line cultivars would have an effect similar to polygenic non-specific or horizontal resistance (Van der Plank 1963) in delaying the intercrop build-up of the pathogen.

The "dirty crop" approach, using partially resistant multi-lines, has a significant potential advantage over the "clean crop" approach, using completely resistant multi-lines (Marshall and Pryor 1978). Since moderately susceptible lines are also considered, the breeder is in an advantageous position as he can exercise selection for other characters like yield, height, maturity etc. (K.S. Gill et al. 1979). It would also indefinitely extend the useful life of strong resistance genes (Van der Plank 1963), including those that have broken down in the past. Hence, it would free the breeder from the difficult task of continually isolating and evaluating new sources of resistance.

However, the proposal by Frey and coworkers is based on the assumption that simple races will dominate the pathogen population and this is based on two questions. First, whether Van der Plank's (1963) axiom that complex pathogen biotypes are less fit than simple biotypes in susceptible host genotypes, and further, whether this loss of fitness can be directly attributed to the virulence genes per se. Second, given that stabilizing selection is a general if not universal phenomenon, whether it is strong enough to prevent the development of complex races of pathogen on multilines, i.e. whether stabilizing selection against races carrying additional virulence genes will outweigh the selection pressure towards greater virulence, which arises from the fact that more complex races can grow on more lines and, hence, spread more rapidly through the host population.

The only common point between the two approaches is the final aim to produce a variety consisting of a number (6 to 15) of phenotypically similar lines which differ in the resistance genes they carry.

The prerequisities of multiline cultivar approach are: proper identification of diverse genetic sources of resistance, adequate race survey, desirable and commercially acceptable and, if possible, a widely adapted recurrent parent (M.V. Rao 1968). By following a conventional or limited back-crossing programme (the latter was found to be more beneficial by Borlaug because of the transgressive segregants, which were better than the recurrent parents), phenotypically similar but genotypically different lines are developed by crossing the recurrent parent with stocks carrying diverse genes for resistance. All these phenotypically similar lines are mixed together and distributed to the farmers as a composite variety. If a line/genotype is affected by a new race, it is withdrawn from the composite bulk.

The following programmes relate to the introduction of multiline cultivars:

The New York Programme. Jensen (1952) proposed the development of multiline cultivars by blending several compatible lines of different genotypes. The pure line thus blended would be non-uniform for disease resistance and agronomic phenotype (Jensen 1966). Such cultivars require judicious selection of components (Grafius 1966, Jensen 1965).

The Rockefeller Foundation Programme. Borlaug and Gibler (1953) used a modified back-cross method described by Borlaug (1959) to develop wheat lines for multiline cultivars. The best recurrent parents and large group of donor parents were chosen. These were crossed and back-crossed. The BC_1 plants were tested with the tester race. The resistant plants almost similar to the recurrent parent for agronomic attributes were back-crossed to the recurrent parent. The process was repeated for three back-crosses. Eight to 16 of the best near-isogenic lines were selected to be blended to form the multiline cultivar.

Borlaug (1965) believed that the development of a multilineal hybrid variety (Fig. 6.7) will offer the best possible means of incorporating a long-lasting rust resistance into a commercial wheat variety. Once an outstanding single-cross hybrid has been developed, it can be converted into a multilineal hybrid. The first step in this process is the development of many lines phenotypically identical with the male sterile female variety A^s (the modified variety A) through a back-crossing programme. Each of these lines, designated male variety A^n, will ge genotypically different from the others in stem rust resistance, and also from that of the male sterile female variety A^s, but none of these lines will possess the genes for restoration of fertility. Therefore, when crossed with the male sterile female variety A^s, the resulting F_1 plants of each cross will be sterile. These male sterile F_1 lines in turn each will be crossed as female with the male variety B^r to provide a heterotic effect, and this second cross will also simultaneously restore fertility to the F_1 commercial hybrid seed formed in the process. The basic commercial variety A and B, both of which are fertile and into which the cytoplasmic sterile and fertility restorer will be incorporated respectively, should also have different stem rust resistance genes.

Each individual F_1 plant of a hybrid produced by this method would be different in its resistance to stem rust and rust epidemics would not be likely develop in such a population.

The first wheat multi-line cultivar was Miramer 63, which was released for commercial production from the Colombian programme (Rockefeller Foundation 1963, 1964). It was developed by crossing Brazilian wheat Frocor with about 600 varieties/lines. More than 1200 lines phenotypically similar but with resistance from 600 parents were developed. Miramer 63, was a mechanical mixture of equal seeds of ten of the best lines with resistance to stripe and stem rust diseases. It yielded more than twice as much as older varieties in some areas. Within 2 years of its release, stem rust was parasitizing on the two component lines; but total losses were always less than the theoretical maximum

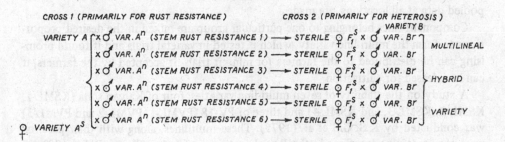

Fig. 6.7. Scheme for development of multilineal weat variety (Borlaug 1965)

of 20%. The two stem rust-susceptible lines and two others were dropped from the multi-line and four new ones from the reserve of over 600 lines were added to form a new multiline cultivar, Miramer 65.

The Iowa Programme. A series of multiline cultivars in oats have been released from a programme in Iowa (Browning et al. 1962, Browning and Frey 1969). They have used two recurrent parents CI 7555 (a Clintland type) and CI 7970 (an early experimental line). Using conventional back-crossing, more than 25 different res genes have been incorporated into the isogenic lines of the recurrent parents. All isolines of oats were tested for agronomic performance in rust-free experiments and for adult-plant reactions to crown-rust races. Isolines equivalent to recurrent cultivar in productivity (i.e., yield, lodging resistance and test weight) and with appropriate crown-rust resistance were composited to form breeder's seed of a multiline cultivar. This seed was used in turn to produce foundation, registered and certified seed (Frey et al. 1971).

These authors used six back-crosses to recover their types. Two series of multiline cultivars have come from the Iowa programme. Multiline E 68, E 69 and E 70 (early) and multiline M68, M 69 and M70 (mid-season) have been developed from this programme. Multiline E 68, contains ten component lines; multiline E 69, eight; multiline E70, eleven; multiline M 68, eight; multiline M69, nine; and multiline M70, seven components.

The Indian Programme. In the recent past, emphasis has been placed on the production of multi-line varieties in the wheat programme in India. Breeders in India have developed multi-line varieties of three promising and popular wheat varieties: Kalyan sona; PV-18, and Sonalika (RR21). These varieties are high-yielding and have good agronomic characters. They were resistant to rusts and other diseases at the time of their release in India, but after 5–6 years of cultivation had fallen susceptible to new races of the pathogen. Therefore, with the aim of replacing these pure line susceptible varieties by their multi-line, this programme was taken up. By this approach, the life of these varieties may be further prolonged. These three varieties have been used as recurrent parent and several donors as non-recurrent parent. The component lines produced by different breeders all over the country are assembled at the main center of the zone, and the main center grows these components along with their recurrent parents at several locations within the zone, i.e. multilocation testing is done and data on several characters such as height, maturity, ear length, disease resistance etc. are recorded at all the locations and based on pooled data component lines which have shown similar behaviour at all the locations assigned to a particular group. In this way, several groups, each comprising component lines which have proven to be the similar on the basis of pooled data at all locations, are made.

Component lines belonging to one particular group are mixed in the desired proportions to form the multi-line variety, which is tested in varietal trials and if found promising can be distributed to the farmers for minikit trials. If accepted by the farmers, it can be released for cultivation.

A study on the stability of seven multi-line varieties; four of Kalyan sona (KSML-1, KSML-2, KSML-3, and KSML-4) and three of PV-18 (PVML-1, PVML-2 and PVML-3) was conducted by K.S. Gill et al. (1975). These multilines, along with their parents, were grown at nine locations in the Punjab province of India. The multi-line differed significantly from each other for heading, maturity, ear length and plant height. They had better resistance than their parents, but showed only marginal increase in yield. KSML-3 and PVML-3 were found to be the most stable multi-lines. Both of these had ten component lines in each. They also reported that the more component lines there are in a multi-line, the more stable will it be. A stable variety is one which has high mean yield, unit regression, but a deviation as small as possible from regression. Based on this, KSML-3 and PVML-3 are found to be the most stable varieties in their groups.

Variation in the multi-lines of PV-18 variety and their components with respect to yield and yield components and reaction to yellow and brown rusts was studied for years. There was considerable variation for these traits among 212 components. Based on agronomic characters and rust reaction, 84 of these were utilized to compose 12 multi-lines of component lines of different height and maturity groups. Some of the multi-lines out-yielded recurrent parent PV-18, whereas others were comparable to PV-18. Multi-lines were better in their resistance and only in the case of PVML-5 (brown rust) and PVML-11 and PVML-13 (yellow rust) was the reaction more than in the component lines (K.S. Gill et al. 1977).

Depending on the reaction types, the components of Kalyan sona and PV-18 multi-lines were classified for seedling and adult plant reactions (K.S. Gill et al. 1978). The first group of component lines of both multi-lines had some susceptibility, both at seedling as well as adult plant stages. The genes for rust reaction thus behaved uniformly at different temperatures. The second group of component lines had resistant seedlings but was susceptible at the adult stage, indicating that the genes conditioning rust reaction were thermo-sensitive. The third group showed seedling-susceptible and adult-resistant reaction. This group also carries thermo-sensitive genes for resistance. Some of the component lines of these multi-lines were resistant both at seedling and adult stages, suggesting that the genes conditioning resistance in these lines were thermo-insensitive.

The variation among the component lines of the KSML-3 (a multi-line based on cultivar Kalyan sona, spring bread wheat) was studied for agronomic characteristics (K.S. Gill et al. 1980). In days to earing and in plant height the variation was small. This helped in improving uniformity to the multi-line. The lines had an improved tillering ability and had larger seeds. This partially explained the increased yield potential of the multi-line as compared with Kalyan sona. All the lines were susceptible to only one or two races of the yellow and brown rust. In no case was any race virulent against all the lines.

6.2.2 Mechanism of Action of Multi-Lines

a) **Reduction of X_0 and r.** The mechanism by which multi-line cultivars probably buffer against disease loss has been described by Borlaug (1959), Browning et al. (1964), Jensen (1952), Simmonds (1962), and Van der Plank (1963). The pathogen increases from initial inoculum x_0 at the rate r in time t, and results in x proportion of susceptible tissue becoming infected (Van der Plank 1963). A variety with vertical resistance (VR), being selectively resistant to the race population, reduces x_0 but r remains unchanged and is usually high (not limiting); x may be very large by the end of the disease season. Therefore, VR is valuable only as long as it gives resistance to all prevalent races and

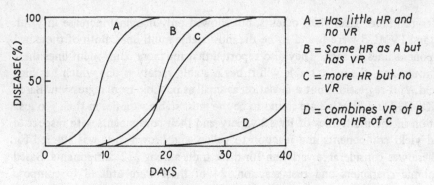

Fig. 6.8. A representation of the effects of VR and HR, separately and combined on disease progress curve (Van der Plank 1968)

keeps x_0 very small for all. A variety with horizontal resistance (HR), i.e. resistant to all races of the pathogen, does not reduce x_0, but reduces r (Van der Plank 1963). Since r is small, the rate of epiphytotic development is reduced to the point where the host matures with small x and little measurable damage.

Van der Plank (1968) represented the effects of vertical and horizontal resistance, separately and combined, on disease progress curves (Fig. 6.8). The HR of variety D has greatly enhanced the variety's VR. Although neither the VR nor HR is very great, as shown by curve B for VR alone and curve C for HR alone, the combined resistances are very effective. Variety D, therefore, becomes diseased only late in the season, i.e. too late for much harm to be done.

The effect of VR and HR on the development of bacterial spot, *Xanthomonas vesicatoria*, in pepper was studied by Dahlbeck et al. (1979). They monitored disease progress in two field experiments in a pepper, *Capsicum annuum* line or cultivar: (1) without known resistance to the pepper strain of *X. vesicatoria*; (2) with HR; (3) with VR, and (4) with the combination of the two types of resistance. Initial infection was caused by the tomato race of *X. campestris* pv. *vesicatoria* in both experiments. In one experiment, VR (alone or in combination with HR) delayed disease on set 1 week. In a second experiment with a lower initial inoculum level, the epidemic was delayed 1 week on pepper with VR and 2 weeks on pepper with VR and HR. In both experiments, HR resulted in reduced rates of disease development and delays in epidemic onset resulted in yield increases. VR with or without HR resulted in significant decreases in disease levels and increases in yield, despite vertical mutability of the pathogen.

As developed, multi-line cultivars utilize high-type racespecific resistance (several genes for VR) and reduces x_0. Such resistances show complete susceptibility to certain races. However, when a series of such resistance genes is placed into a multiline cultivar, the population of plants making up the cultivar has degrees of resistance to all races. The reduced rate of rust build-up in the multiline cultivar was accomplished by the resistant plants serving as "spore trap" that kept potential parental spores from contributing offspring in future cycles. Typical seasonal spore productivities on pure-line and multi-line oat cultivars inoculated with races 1 and 6 of crown rust are shown in Fig. 6.9, which shows an epiphytotic being delayed in multi-line as compared to pure line. As-

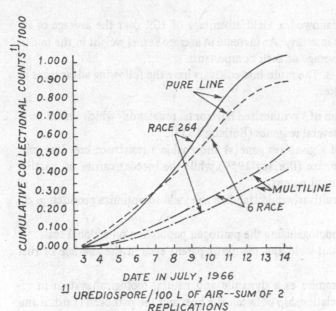

Fig. 6.9. Cumulative daily counts from spore collectors outside 16 × 16 m plots of oat crown rust uredio-spores per 100 litres of air, expressed as proportions of 1000 and plotted against time for the 4 treatment combinations (adapted from Frey and Browning 1969)

suming a viable spore of a race virulent on two of the ten components of a multi-line, the probability is only 0.2 that the plant will be susceptible. If the spore lands on a susceptible plant, infects the tissue and forms a pustule, the progeny have a probability of 0.2 that surrounding plants will be compatible host (Browning and Frey 1969). Thus not only is χ_0 reduced as the characteristic of cultivar with VR, but r also is reduced, as is characteristic of HR. Browning and Frey (1969) have called this "synthetic horizontal resistance" or population resistance.

Luthra and M.V. Rao (1979) studied the mechanism of action of multi-line cultivars in wheat against leaf rust. Seven experimental multi-lines with 0, 28, 40, 50, 58, 60, and 70% susceptibility were subjected to leaf-rust epiphytotics in the field along with their pure line components. Both initial inoculum (χ_0) and rate of increase (r) of leaf rust were substantially reduced in the multi-line cultivar. Initial inoculum (χ_0) was reduced by 45 to 75% and the overall infection rate (r) by as much as 16% over the average of components. As a result of reduced χ_0 and r, the intensity of leaf rust in the multi-lines was also significantly affected at all stages of rust development. It was reduced from 32.1 to 89.54% over the average of components differing from one multi-line to another and also from time to time. The susceptible recurrent parent, Kalyan-sona at the peak of rust infection exhibited 86.75% severity, which in the multi-lines it ranged from 5.8 to 35.0%. The rate of increase (r) in the multi-lines was proportional to the logarithm of the proportion of susceptible plants in the host mixture.

They noted that even if as many as 50% susceptible plants are present in a multi-line they would not suffer much loss from rust damage. They further found that due to lower χ_0 and r, the multi-lines help to delay the onset of epidemic. Leaf rust epidemic took 31 days to reach the peak in the susceptible variety Kalyan sona, the multi-line like MLKS-7506 took about 49 to 50 days to reach the same level of infection. The multi-lines, therefore, appeared to have the benefit of both VR and HR.

Some of the multi-lines showed a yield advantage of 10% over the average of its components, due to low rust intensity. An increase in average kernel weight in the multi-line was obtained over the average of their components.

Advantages of Multi-Lines. The multi-line cultivars have the following advantages as compared to pure line varieties:

— They provide a mechanism of "synthesized horizontal resistance" which unlike pure line cultivars can utilize several resgenes (Borlaug 1959).
— They extends the life of a given res gene(s) and enable a resistance breeding pro-gramme to be reduced in size (Borlaug 1959) while the breeder carries on parallel improvement in the recurrent parent.
— The use of a multi-line cultivar would stabilize the yield to optimize production on a given farm.
— This reduces the risk of homogenizing the pathogen population on a global scale.
— Multi-line cultivar may out-yield the recurrent parent even in the absence of rust infection.
— Multi-lines hold great promise as a dynamic and natural biological system in ef-fectively balancing the relationship between the host and the pathogen (Luthra and M.V. Rao 1980).

Criticisms of the Multi-Line Approach. The multi-line approach of management of VR genes for controlling pathogens with great epiphytotic potential such as rusts has been criticized as expensive (Suneson 1960), agronomically conservative (R.M. Caldwell 1966 and Hooker 1967a), a breeding ground for new races (Hooker 1967, Simmonds 1962, Van der Plank 1963), and possibly even a super-race (R.M. Caldwell 1966).

6.2.3 Multiple Resistance

Resistance to two or more diseases has been bred into individual cultivars since. W.A. Orton (1909) succeeded in combining resistance to *Fusarium* wilt and root knot nematode in cowpea and cotton. Hope and H44 wheat cultivars combined resistance to leaf rust, stem rust and covered smut (Ausemus 1943). Multiple resistance has long been a breeding achievement in tobacco, sugarbeet, corn, beans, and many other crops. Some noteworthy accomplishments have been reported recently for cabbage (P.H. Wil-liams et al. 1968), cucumber (W.C. Barnes 1961, Sitterly 1972), sugarbeet (Gaskill et al. 1970) and tomato (J.P. Crill et al. 1971). Resistance in wheat to both cereal leaf beetle, *Oulema melanopus* and to stem sawfly, *Cephus cinctus* were incorporated by L.E. Wallace et al. (1974). Multiple-pest-resistant alfalfa lines were developed by Kindler and Schalk (1975) and Peaden et al. (1976).

Multiple resistance in induced amphiploids of *Zinnia elegans* and *Z. angustifolia* to three major pathogens was reported by Terry-Lewandowski and Stimart (1983). The amphiploids possessed high levels of resistance to powdery mildew, *Alternaria* blight and moderate to high levels of resistance to bacterial leaf and flower spot.

One of the approaches for incorporating the multiple resistance could be the screen-ing for multiple resistance line(s) from germplasm followed by crossing and screening for more than two diseases and/or insect pests in the segregating generations. Such a technique is being followed in cowpea. Five thousand cowpea lines were screened for field resistance to anthracnose, *Cercospora* leaf spot, rust and bacterial pustule. After preliminary observations on 5000 lines, 719 lines were selected for further evaluation in field nurseries. All 719 lines were resistant to at least one of the four diseases, 685 lines were resistant to at least two diseases, 537 lines were resistant to at least three diseases, 208 lines were also resistant to all four diseases and of these 28 lines were also resistant to target spot (R.J. Williams 1977).

Another approach could be followed by making single-cross F_1 hybrids. These F_1 hybrids could be crossed with other resistant donors to make double or top crosses to combine resistance to given diseases and insects. If more donors are available for a single parasite, it is desirable to use good combiners for yield components and plant type etc. The pedigree method is suitable for handling the segregating generations (Khush 1977a). At the International Rice Research Institute (IRRI) in the Philippines efforts have been made to eliminate the susceptible material in the early generations. The screening starts even in the F_1 generation of a cross. Khush (1977a) elaborated the procedure citing an example of a double cross between four parents of which A, B, C, and D are resistant to bacterial blight, grassy stunt, brown planthopper, and green leafhopper, respectively. All of these traits are monogenic and dominant in their inheritance and are inherited independently. About 400 seeds from the double cross (A/B//C/D) are obtained. These seeds are germinated and inoculated with grassy stunt virus in greenhouse. Approximately 50% of the seedlings are susceptible and are, therefore, eliminated. The remaining 200 are transplanted in the field and inoculated with bacterial blight. About 50% of these plants are susceptible and are rooted out. The remaining 100 plants are harvested separately. Two small seed samples are taken from each, and the progeny are tested for resistance to brown planthopper and green leafhopper. Those carrying the brown planthopper resistance gene (50%) and those carrying the green leafhopper resistance gene (50%) are identified. The F_2 populations are grown only from those carrying both the genes (25–30 plants), which segregate for the four resistance genes. The population is subjected for appropriate disease and insect pressures and agronomically desirable plants with multiple resistance are selected in F_3 and F_4 generations to obtain true breeding lines. IR 28, IR 29 and IR 34 are some of the varieties of rice with multiple resistance developed at IRRI.

6.3 Management of Tolerance

A tolerant cultivar is one which endures disease attack (R.M. Caldwell et al. 1958) and looks susceptible but shows significantly smaller yield reductions than a susceptible one (Browning and Frey 1969).

Gokhale and Patel (1952) reported that with equal amount of stem rust infection the percentage loss in grain weight of certain wheat varieties was about 95%, while in tolerant varieties it varied from 10 to 30%. Hayden (1956) reported that Lee and Sentry wheats were tolerant to *P. graminis* f.sp. *tritici* race 15B when compared to Marquis, Carleton and other varieties. Heavy versus light inoculations made relatively little difference in Marquis and Carleton but affected Lee and Sentry. Furthermore, the rate of

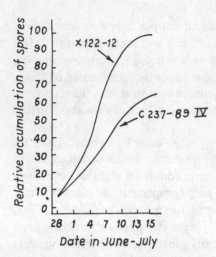

Fig. 6.10. Relative cumulative counts of crown-rust spores collected outside of plots sown to susceptible (x 122-12) and tolerant (C237-89IV) oat lines that had been inoculated with 4 races (adapted from Frey et al. 1979)

rust spread was slower on Lee and Sentry. Van der Plank (1960) interpreted the tolerance of these cultivars as partial resistance manifested not by lower infection type, but by a slower rate of increase. Torres and Browning (1968) and Cournoyer (1970) reported that tolerant cultivars reduced spore production, and it would take less photosynthate from the oat plant and thus more of the photosynthate would be available for filling the caryopsis. The reduced spore production on a tolerant cultivar than on susceptible one is shown in Fig. 6.10.

Frey et al. (1979) noted that oat plants that are tolerant to crown rust are judged susceptible on the regular rating scale, but when mature, rusted tolerant plants produce higher grain yields than do rusted plants of a non-tolerant cultivars. They therefore suggested that degrees of tolerance must be assayed on the basis of yield reduction due to rust attack. Simons (1965) and Simons and Browning (1961) developed a technique for indexing oat genotypes for tolerance via the ratio of productiveness in paired plots of oats of a cultivar, one with rust and one without rust. The tolerance index for an oat line is:

Tolerance index = grain yield in plot with rust/grain yield in plot without rust.

In oats, seed weight is the only yield component that can be affected by rust disease, therefore Frey et al. (1979) substituted 200 seed weight for grain yield in the tolerance rating technique. Typical tolerance indexes in some oat cultivars are presented in Table 6.1. Cherokee has been known for 25 years for its tolerance. Benton is more tolerant to crown rust than its sister line Clinton (R.M. Caldwell et al. 1958). The landhafer gene, confers resistance to races 203 and 216 in Clintland, has confounded the tolerance rating with race 290, a race virulent on Landhafer; the Clintland tolerance index reverts to that of its isoline, Clinton. This suggested that selection for tolerance to rust must be done in the absence of genes for vertical resistance because if VR gene(s) are present, can confound the tolerance indexes. This can be done either (1) by eliminating VR genes from the populations before initiating selection for tolerance or (2) by the use of rust races which may be virulent on all the genotypes in the plant populations.

Table 6.1. Rust reaction and tolerance indexes of four oat cultivars to different races or mixtures of races of crown rust. (Simons 1966)

Cultivar	Rust race			
	203 and 216		290	
	Reaction	Tolerance	Reaction	Tolerance
Clintland	R	0.86	S	0.67
Cherokee	S	0.81	S	0.83
Benton	S	0.74	S	0.75
Clinton	S	0.65	S	0.70

R = Resistant, and S = susceptible.

The inheritance of tolerance has been discussed in Chap. 2. Sources of genes for tolerance of oats to *Puccinia coronata* f.sp. *avenae* have been found in cultivated types by R.M. Caldwell et al. (1958), Simons (1968) and Simons and Murphy (1967) and in wild types of *Avena sterilis* by Simons (1972b). Tolerance has also been observed in mutagen-derived lines of oats (Simons 1971).

6.4 Management of Horizontal Resistance

Breeders and pathologists have preferred to utilize the specific (VR) resistance in the past, because of the ease with which it can be incorporated into varieties, due to its simple inheritance (often a single gene) and easily recognizable character (R.R. Nelson 1975). However, with few exceptions, it is generally acknowledged that VR is unstable. The pathogen usually requires a single genetic change to overcome a single gene for resistance (Van der Plank 1968, R.R. Nelson 1973, 1975). Because of this, many resistant varieties have had short-lived usefulness (J.D. Hayes 1973).

Horizontal resistance (HR) provides an alternative to VR which is evenly spread against all races of pathogen (Van der Plank 1963). Its stability is postulated to be due to its polygenic inheritance. It is also presumed that HR acts to reduce the effectiveness of one or more components of parasite fitness in the pathogen, through accumulation of resistance genes which act quantitatively in the host (Van der Plank 1968). R.R. Nelson (1973, 1978) did not agree with this interpretation and therefore redefined HR as resistance that reduces the apparent infection rate (R.R. Nelson 1978). Field resistance or generalized resistance or adult plant resistance etc, have been used in many contexts as alternative to non-specific (HR) resistance. The mechanism of action and breeding of HR is discussed in the foregoing paragraphs.

a) Components of HR. The non-specific resistance or HR to rust may involve exclusion of fungus (Romig and R.M. Caldwell 1964); limitation of pustule size (Krull et al. 1965), or possibly slow growth and development of the fungus. The joint action of these host characters may drastically slow down a disease epidemic to the point of insignificance (Van der Plank 1963). Characters such as these, involved in slow rusting, were observed by many, including Farrar (1898) and R.M. Caldwell et al. (1957) with leaf rust *(P. recondita)* and Hooker (1967) with maize rust *(P. sorghi)*.

In stem rust of wheat, slow-rusting lines were observed to have a low apparent rate of infection compared to susceptible varieties (MacKenzie 1976). The components of rust development in slow rusting in barley infected with *P. hordei* were related to each other and to the area under disease progress curve, as indicated by significant correlation coefficients (D.A. Johnson and Wilcoxson 1978). Slow leaf-rusting varieties of wheat had longer latent period and restricted pustule size (D.A. Johnson and Wilcoxson 1979, Kuhn et al. 1978, Ohm and Shaner 1976, Shaner et al. 1978, Statler et al. 1977). Kochman and Brown (1975) and Luke et al. (1972) also found slow rusting in oats, where the increase in the percentage of crown rust was slower. Slow mildewing in wheat and barley has also been reported (Parlevliet 1976, Parlevliet and Ommeren 1975, Shaner 1973a,b,c).

Rouse et al. (1980) studied sporulation capacity (SC) and infection efficiency (IE) of six isolates of *E. graminis* f.sp. *tritici* on winter wheat cultivars. The change in ranking and the significant statistical interaction in the analysis of variance of isolates by cultivars with respect to SC and IE demonstrates genetic variability for these traits in wheat and *E. graminis* and indicates the possibility for erosion of slow mildewing resistance.

The effect of cultivar and development stage on infection frequency (IF) was studied in the seedling and adult plant stage of 15 spring barley cultivars against leaf rust, *P. hordei* (Parlevliet and Kuiper 1977). In both stages the cultivar effects were highly significant. The cultivars L94 and Vada represented the extremes, Vada having about 2.5 times fewer uredosori than L94. Between cultivars and development stage, clear interactions occurred. Pauline f.i. had the same low IF as Vada in the seedling stage, but in the adult plant stage its IF was about 70% higher. Also other effects could influence the cultivar effects. Increasing leaf age appears to increase IF. The cultivar effect also seems to depend on the level of IF. At high levels the cultivars differed far less than at low levels of IF. The cultivar effect on IF appeared correlated with partial resistance in the field ($r = 0.7$) through a high correlation with the cultivar effect on latent period ($r = 0.8$).

I.T. Jones (1978) studied the components of adult plant resistance to powdery mildew, *Erysiphe graminis* f.sp. *avenae*, in four oat genotypes. A negative relationship existed in these genotypes between length of latent period and the percentage leaf area with mildew. The longer latent period of Cc 4761, the highly resistant cultivar, was reflected in the delay showed by this genotype in reaching its relatively low peak production of spores. The latent period can be used as a factor for selecting the adult plant resistance in early growth stages.

The level of partial resistance to leaf rust, *P. hordei* in West-European barley was measured in 40 West-European spring barley species by Parlevliet et al. (1980). The cultivars varied widely for partial resistance, many of the cultivars carrying a considerable level. Both latent period and infection frequency showed large differences between cultivars. The latent period in the adult plant stage being correlated ($r = 0.82$) with partial resistance, and infection frequency in the seedling stage only rather weakly

($r = -0.33$). Selection for partial resistance appeared very effective in all stages tested. Selection in the small plot stage was most effective, followed by selection in the seedling stage.

If components of HR are independent and under separate genetic control, it may be possible to recombine them to increase the level of resistance. The smaller pustule size and latent period component of slow rusting in the wheat for leaf rust are governed by two genes. It suggested the possibility of selecting for these components in segregating populations (Kuhn et al. 1980), and the host genotypes with more stable resistance by the assembly of these components may be developed.

Recognizing that the contribution of a component of slow rusting to an epidemic cannot be determined by a simple measurement of its magnitude in monocyclic infection experiments. Shaner and Hess (1978) developed equations to predict the effect of any given set of resistance components on the curve of disease progress in the field. Using their equations Kulkarni et al. (1982) measured the relative importance of different variables determining the course of an epidemic. Only four components, namely, latent period, infectibility, sporulation and weather were found to be equally important and they collectively determined the rate of progress of leaf rust *(P. recondita)* in wheat.

Components of partial resistance of wheat seedlings to *Septoria nodorum* were measured in the glasshouse by Jeger et al. (1983). There was no clear evidence of associated variation in components. Pathogen-induced necrosis, related possibly to toxin susceptibility, and unit spore production, were major components of partial resistance to *S. nodorum*.

b) Heritability of Field Resistance. Simons (1975) reported heritability of field resistance to the oat crown rust fungus. Four unadapted strains of oats, *Avena sativa*, with field resistance to crown rust were crossed with an unadapted but susceptible cultivar. Lines from each of the four crosses showed continuous variation from susceptibility to resistance characteristic of polygenic inheritance. Heritability values estimated from components of variance for resistance, measured in terms of yield reduction, ranged from 46 to 86% and in terms of reduction in kernel weight from 65 to 92%. The relationship of yield to resistance in the absence of rust was generally negative, with correlation coefficients ranging from 0 to -0.69. None of the lines tested combined maximum yield with maximum resistance.

Kuhn et al. (1980) reported the heritability for the components of slow leaf rusting against *P. recondita* in wheat. The higher heritability of the latent period (49%) suggests that selection based on it could be more successful than selection based on pustule size, which had a heritability value of 27% only. Shaner et al. (1978) suggested that for breeding work this could be done relatively easily by making observations, one at 8 or 9 days after inoculation to evaluate the latent period and one at 14 days to measure pustule size and infection frequency.

c) Breeding of HR. The breakdown of disease-resistant varieties with true resistance alone leads breeders to re-evaluate field resistance. The methods of estimation may be by (1) developing isogenic lines with known levels of field resistance with that of breeding materials or newly developed varieties to be tested; and (2) by elucidating the nature of multiplication of the pathogen on the varieties to be tested and by estimating infection rate, a measure of field susceptibility which is an irony of field resistance.

In a seminar on HR to rice blast disease, Vanderplank (1975) made six suggestions as follows:

− HR may be determined as field resistance. In the absence of VR, resistance is HR. He suggested that if one can exclude all VR genes, one can simply compare cultivars or lines in the field, and the comparison will measure HR alone.
  The method then is to expose lines (or cultivars) to infection in the field by virulent races to which the lines are (vertically) susceptible. The resistance that remains is horizontal. The difficulty arises here that the lines must be exposed to races virulent on all of them or to a mixture of several races, some of them virulent on some lines but avirulent on others (then VR enters and confuses the results).
− The lines/varieties that are more difficult to infect, e.g. those that develop the fewest lesions per plant, per leaf, or per cm^2 of leaf may be selected.
− The lines/varieties in which period from inoculation to sporulation is longer may be selected.
− The lines/varieties in which sporulation is less abundant may be selected.
− HR may be enhanced by breeding.
− The HR and VR may be combined.

Yorinori and Thurston (1975) and Rodriguez and Galvez (1975) attempted what Vanderplank suggested. They observed that the resistant type of lesions were smaller, generally darker in colour, took longer to sporulate and produced fewer spores than the susceptible types. They also observed diversity in lesion types even with the use of single isolates on the varieties and selections used. In addition, vertical interaction between the isolates (races) and the host variety was noted. However, Yorinori and Thurston (1975) did not offer any definite conclusion regarding the general resistance of rice to *Pyricularia oryzae*. This was due to the inconsistency in their results.

The selection of HR by measuring the apparent infection rate (r-value) of both the leaf and neck blast phase of the rice blast disease has been tried by the investigators at the Institut de Recherches Agronomiques Tropicales et des cultures Vivieras (IRAT) at the Ivory Coast, West Africa (IRAT 1973). Low r-values were obtained for leaf blast in varieties, Moroberekan (0.045), Kataktara (0.060), Fassa (0.065), and Te-tep (0.073). For neck blast infection, the apparent infection rate (r) was calculated for 30 varieties and ranged from 0.115 to 0.692, the average being 0.258 ± 0.142 (IRAT 1977). Similarly, using the data of Van der Plank (1963, 1968), they were apparently able to identify certain varieties exhibiting either a small number of lesions or a long latent period (IRAT 1976).

The rate of lesion development of blast disease on the resistant varieties Te-tep, Dawn and Sensho was very slow, so that 60 days after sowing, total disease was much less when compared to the susceptible varieties Fanny Khao-teh haeng, Peta and IR 442-2-58 (IRRI 1970, 1978) (Fig. 6.11). Disease efficiency (DE), latent period (LP), lession size (LS) and sporulation capacity (SC) of three isolates of *P. oryzae* (Cav.) were measured on rice varieties Te-tep, IR 442-2-58, Gogowierie, Daurado Precose, IRAT 13 and 1021 (Villareal 1980). The overall analysis of the results shows that these varieties possess different levels of HR. High levels of HR were clearly demonstrated in IRAT 13, and Gogowierie. It was also shown that the major rate-reducing factors among these varieties were their capacities to reduce DE, LP, and SC with the exception of variety 1021,

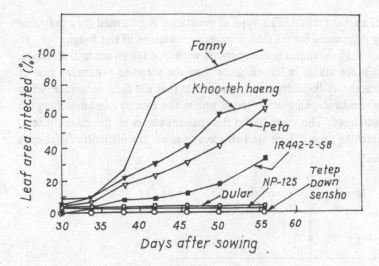

Fig. 6.11. HR to blast disease of rice was confirmed for Sensho, Dawn, Tetep, Dular and NP-125 when they developed almost no infection being inoculated with isolates of blast from both their own leaf lesions and those from seven other varieties (adapted from IRRI 1978)

which had the ability to reduce DE and SC, but had an LP that was comparable to that of the susceptible check. The results support the assumption that slow leaf blast infection is a form of HR.

Hagborg (1979) gave a method of estimating field resistance to black chaff (*X. campestris* pv. *undulosum*) in wheat. Artificial epiphytotics were induced in replicated plots of a large number of varieties in successive years. Maintenance of the same control variety, Marquis, throughout all tests permitted a comparison with it under different environmental conditions in different years. Mean rating for Marquis was taken as the numerator and the mean rating of the test variety in the same years as the denominator. The quotient represented the resistance of the variety when Marquis was unity.

Parlevliet (1983) noted that if the host population varies for both HR and VR, selection for HR requires that the two types can be distinguished. This is not always easy. Using a mixture of races makes it more difficult, if not impossible. Using a single race provides the best combinations for the selection of HR in the presence of VR. This race should have the broadest possible virulence spectrum to suppress the expression of as many VR genes as possible.

d) **Breeding of Field Resistance with True Resistance.** The severe breakdown of true resistance for rice blast observed in the varieties Kusabue, Yuukara, Pi5 and others is undoubtedly caused by lack of field resistance that results from the Vertifolia effect. True resistance itself appears to have no or little effect on field resistance (Asaga and Yoshimura 1969). Toriyama (1975) stated that field resistance was equivalent to horizontal resistance. The break-down of field resistance was conspicuous in Chugoku 31. The resistance of this variety was governed by a single major gene Pi-f; however, its expression was quantitative. The moderate level of field resistance is generally accepted as race non-specific, i.e. horizontal. This type of resistance is governed by a polygene system and is hardly influenced by the shift of races or mutation of the fungi.

If a useful level of field resistance is incorporated with true resistance into a variety, that variety will be more stable in its resistance than the varieties possessing true of field resistance separately. A difficulty arises, however, in that the field resistance, when combined with true resistance, cannot be tested unless the race or races matching the true resistance are obtained. The more useful the true resistance is, the more difficult it is to obtain the matching races. There are two ways to solve this difficulty (Kiyosawa et al. 1975):

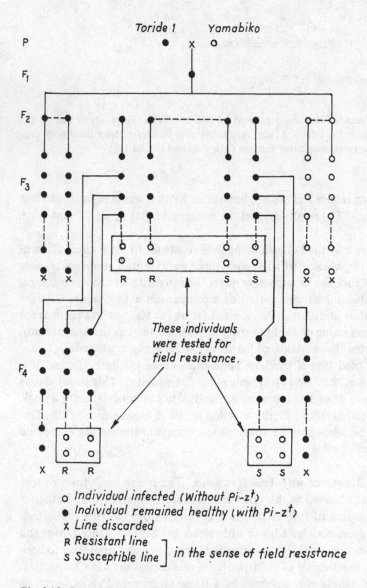

Fig. 6.12. Procedure for evaluating the field resistance of the progenies from the cross between Toride 1 and Yamabiko (Asaga and Higashi 1973)

- To select a fungus mutant matching true resistance; and to use it for the inoculation in the glasshouse.
- The other method, developed by Asaga and Higashi (1973), is an indirect evaluation method (Fig. 6.12). They attempted to incorporate the field resistance of Yamabiko with the true resistance of Toride 1; Yamabiko has Pi-a alone as a true resistance gene for blast disease and is infected by most of the fungus races in Japan, but has rather high field resistance. Toride 1 has the true resistance gene Pi-z^t introduced from an Indian variety, TKM 1, which is operative against all the fungus races in Japan. The individuals without Pi-z^t were selected from the F$_3$ lines that segregated according to the resistance due to Pi-z^t, and tested for field resistance by the ordinary method in the glasshouse or in the field. From the results, Asaga and Higashi estimated the field resistance of the remainder of the same lines, that is, individuals with Pi-z^t. They tested the F$_4$ plants again by the same method. They developed eight lines that combined the true resistance gene Pi-z^t from Toride 1 with field resistance as high as that of Yamabiko.

7 Production of Disease-Resistant Plants by Unconventional Breeding

Classical as well as unconventional plant breeding programmes consist of the production of variability in the plant population followed by the selection of desirable types (Wenzel 1985). Long before man began to influence plant breeding, the evolutionary selection of resistant types took place, resulting in land races, and when Mendelian laws were rediscovered 80 years ago, breeding for resistance continued as a major approach to increasing total yield (Borlaug 1983).

Plants regenerated from cell/tissue cultures are frequently different for one or more characteristics from the parent plants which originally gave rise to these cultures. Nabors (1976) has described and assessed the relative merits of using either spontaneously occurring or induced mutants in cell cultures, to obtain agriculturally useful plants. He concluded that if little is known about the inheritance of the desired trait, it is better to search for spontaneous mutants in tissue cultures first, and if this fails one may attempt to induce mutations. It would be advantageous to use haploid cell lines for selecting recessive mutants.

The most important advantages of in vitro cell and tissue culture in breeding are: (a) freedom from the effects of climate and the natural environments, which make it easier to measure slight quantitative differences in polygenically inherited resistances, (b) the ability to handle large number of individuals in a very small place; and (c) the ability to work with microspores and haploids, the simpler genomes of which allow the uncovering of recessive traits and additive characters within a relatively small population. These combined advantages speed up the production of a new variety (Wenzel 1985). However, unconventional techniques can never fully replace or be used independently of classical breeding, but may be used as complementary to the overall breeding process.

There are some major problems confronting the plant breeder who wishes to utilize selection at the tissue/cell culture level for crop improvement. Firstly, not all cultures regenerate whole plants readily. Sometimes, this difficulty can be bypassed by utilizing embryo culture, but in general, whole plant regeneration from cereal and legume tissue cultures is not readily possible when small segregates of cells and single protoplasts are utilized (Bhojwani et al. 1977; King et al. 1978). Therefore more basic work is needed to develop simple techniques for reproducible large-scale regeneration of plantlets. Secondly, whether traits selected at the cell or callus level will be expressed in the whole plants regenerated from them and whether these traits will be transmitted to the subsequent generations cannot always be predicted with an acceptable degree of confidence.

7.1 Basic Techniques in Plant Cell Culture

a) **Callus and Suspension Cultures.** In most laboratories few basic media with modifications are required to optimize growth of specific plant cells. These media consist of inorganic salts, sucrose as carbon source, thiamine and inositol. Almost all plant cells require hormones, an auxin and a cytokinin for growth. They may or may not require vitamins, such as nicotinic acid and pyridoxine, for optimal growth.

The primary type of growth that one usually encounters on placement of plant tissues (stem, leaves, petals, hypocotyl sections, apical meristems, anther tissue or embryos) on to a synthetic medium is the callus. The callus, as defined by Dougall (1972), is "cells grown as a coherent mass on the surface of a solidified medium". To promote the formation of callus, an explant of the tissue is aseptically removed from the plant and placed in contact with a medium which contains the necessary callus-promoting substances. The resulting growth may consist of a single-cell type, or more generally of several cell types.

After callus is established, it is usually transferred from solid to liquid medium to establish a suspension culture of cells. Alternatively, the original explant may be placed in liquid medium to obtain a suspension culture and very small groups of cells. A suspension culture may be filtered through appropriate sieves to obtain largely single cells and groups of a few cells in the filtrate. Single-cell clones may be isolated by suspending single cells in a drop of liquid medium or by spreading them on agar medium in Petri dishes.

Quantitatively inherited mechanisms of black shank (*Phytophthora parasitica* var. *nicotianae*) resistance were expressed in tobacco callus cultures; an in vitro host-pathogen system may be useful in screening tobacco *(Nicotiana)* lines for black shank resistance (Deaton et al. 1984).

b) **Haploid Culture from Pollen.** The induction and culture of haploid plants from pollen holds a great potential because at the monoploid level the difficulties due to dominance are eliminated and most of the genes may be expressed. This allows isolation of induced recessive mutations. If the chromosome number of a haploid (which are sterile and do not set seed) is doubled, a completely homozygous and fertile strain is produced. According to Nitsch (1972) only two critical factors are important for plantlet initiation: first, the stage of pollen development, and second, the culture media. The critical stage is just as or before the microspore nucleus divides into two nuclei, and the only nutrient requirements are sugar, iron and mineral salts. Using this technique, it is possible to establish a suspension culture of haploid cells and manipulate them in a manner similar to that of diploid cells.

c) **Protoplast Isolation and Culture.** Protoplast is the "portion of the cell remaining after the cell wall has been removed by enzymatic or physical means" (Cocking 1972). Carlson (1973) has discussed the reasons for using protoplasts in the genetic experiments. These include (a) the potential to obtain a large, homogneous population of cells of known genetic composition from a single plant, (b) the absence of cell wall permits protoplasts to fuse with one another and allows uptake of exogenously supplied particles, (c) plating efficiencies much higher than with intact cells are possible, and

(d) the possibility exists of regenerating the entire organism from a single protoplast. The major problem is one of inducing protoplasts, after various genetic manipulations, to grow back a cell wall, begin dividing and ultimately regenrate the entire plant.

Protoplast-derived calluses of tobacco (*Nicotiana tabacum* cv. Samsum) were selected for their resistance to toxins from *Pseudomonas syringae* pv. *tabaci*, which causes wildfire disease, and from *Alternaria alternata* pathotype tobacco which causes brown spot. A number of plants were regenerated from each of the toxin-selected protoplast-derived cultures. A large percentage of plants obtained from the second selection cycle calluses were resistant to infection by these pathogens. Resistance to wildfire disease, however, seems to be unrelated to resistance to brown spot disease. Variations in the morphological characteristics of the regenerated plants were found. Results of an array of the R_1 generation indicate that the resistance shown by R_0 plants against both diseases is heritable (Thanutong et al. 1984).

d) Totipotency and Embryogenesis in Cell Cultures.

One of the most·unique features of some plant cells is their potential to differentiate into whole plants which exhibit the same phenotype as the parent plant from which the cell was derived. Ohyama and Nitsch (1972) demonstrated that intact tobacco plants could be regenerated from protoplasts, clearly proving the totipotency of plant cells and protoplasts. This process is clearly under hormonal control with considerable environmental infiuence, and the ability to stimulate embryogenesis in somatic and gametic cells represents one of the most powerful genetic tools and perhaps the most critical step in applying plant cell culture techniques to plant breeding (Bottino 1975).

The unconventional approaches which can be useful in the isolation of disease resistance plants are presented in Fig. 7.1 and are discussed below.

e) Selection of Mutations from Pathotoxin-Resistant Cells and Clones.

The primary sources of genetic variation in living systems are mutation and recombination. Recombination is routinely exploited in breeding and much work has recently been directed to increasing the level of mutation in higher plants. According to Bottino (1975), the problem at the cellular level is not one of induction, but of selection. However, recently it has been shown that mutations can be selected in higher plants at the cellular level. The selective technique consists of growing cells which may or may not be treated with mutagens on a medium which contains a toxic substance at a concentration which will normally inhibit growth of all the cells. The cells which proliferate on such a medium would be considered presumptive mutants. The proliferating cells may redifferentiate into plants, and the whole plant may show the mutant phenotype. However, some of the published reports suggested no relationship between the reaction of the plant against pathogens in tissue culture and the reaction of the same genotype as a plant or vice versa. Keen and Horsch (1972), based on their work with soybean and *Phytophthora megasperma*, warned against the use of unnatural host-pathogen systems for screening.

The technique enables the screening of a large number of mutagen-treated cells in excess of 10^6 in a Petri dish, whereas a large field would be required to screen 10^3 diploid plants. This approach may be used for any plant disease, provided the toxin produced by the pathogen concerned is directly available, or a compound related to the toxin has been identified and can be used as the selective agents.

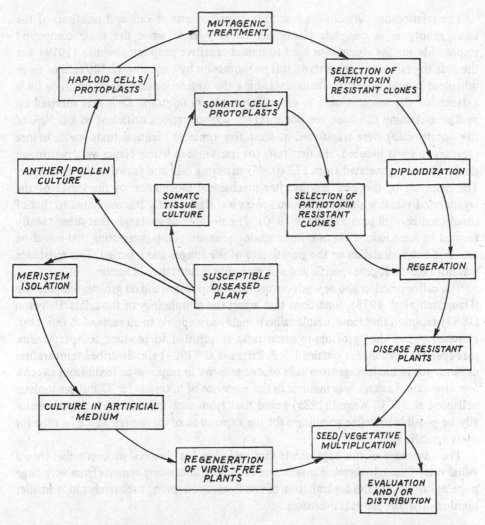

Fig. 7.1. A diagrammatic scheme for obtaining disease resistant or disease free plants by tissue culture

Bajaj and Saettler (1970) observed that bean (*Phaseolus vulgaris* cv. *Manitou*) callus cultures showed differential growth response when cultured on a host-specific toxin filtrate from the halo-blight bacterium, *Pseudomonas phaseolicola*. There was inhibition in growth to the extent of 77%; however, some islets of cells resumed growth, showing differential tolerance to the toxin filtrate. The maize pollen from sensitive or resistant plants can be screened against *Drechslera maydis* pathotoxin. Pollen from resistant plants will germinate at levels of the toxin which inhibit germination of pollen from susceptible plants (Laughnan and Gabay 1973). Differential response of protoplasts isolated from resistant and susceptible plants of maize to the host-specific race T toxin produced by *D. maydis* was demonstrated (Pelcher et al. 1975). It was further demonstrated that callus from susceptible plants (male sterile) showed sensitivity to the T toxin medium, whereas callus from resistant cell lines resulted in the regeneration of complete plants which were resistant to *D. maydis*.

The relationship between the reactions of protoplasts or calli and reactions of the same genotypes as complete plants can be very close when the toxic compound responsible for the virulence is used to detect sensitive proplasts. Behnke (1979) was the first to perform experiments on this phenomenon by screening 42,200 calli of three dihaploid potato clones on media containing the unfractionated liquid or crude filter extracts of the medium used to grow *Phytophthora infestans*. Calli that survived on media containing the toxic compound(s) (in concentrations sufficient to kill 90% of the potato calli) were transferred at least five times to identical toxic media before regeneration was induced. His first tests for resistance in whole plants were performed on 34 clones regenerated from 173 (0.4%) surviving calli and grown in the glasshouse. The diameter of the local lesions after mechanical inoculation of the leaves of the regenerated plants with *Phytophthora* spores was significantly less compared to that of unselected control plants (Behnke 1980). The number of sporangia that subsequently formed in local lesions did not differ among resistant plants, suggesting that infection rate but not sporulation or the growth rate of the fungus was retarded. This resistance was not race- or R-gene specific and was probably quantitative in nature.

The callus reaction also depends on medium composition and on growth temperature (Haberlach et al. 1978), conditions that affect the morphology of the callus. Helgeson (1983) reported that loose, friable callus is highly susceptible to all races of *P. infestans*. However, when the cytokinin-to-auxin ratio is adjusted to produce compact callus, race-specific resistance is obtained. S.P. Briggs et al. (1984) also described temperature dependence in ion leakage from cells of *Avena sativa* in response to toxin from *Drechslera victoriae*. Leakage was minimal in the presence of toxins at 12 °C, but protoplasts collapsed at 35 °C. Wenzel (1985) noted that from such information it might eventually be possible to define conditions for the expression of disease resistance in vitro for every specific system.

The advantage of this approach is that selection is a chemical process rather than a visual one. Through chemical means one can select interesting variants from very large populations of cells and the limitation of visual selection among a relatively much smaller number of whole plants is overcome.

f) Regeneration of Plants from Somaclonal/Protoclonal Variation. New characters are expected from a homogeneous population (homozygous or heterozygous) without meiotic recombination. Organs, tissues, or cells from a given material are cultured in vitro and then regenerated to plants (Wenzel 1985). The objective is to find spontaneous somaclonal variation or protoclonal (from protoplasts) variation among the regenerated plant population, i.e. mutations.

Genetic erosion is a common trait of plant tissue cultures. On prolonged sub-culturing, they undergo changes and show various nuclear and chromosomal conditions, mutation and ploidy levels, which lead to the formation of genetic mosaics in tissue cultures. The naturally occurring chimera with desirable mosaics, that is with resistance, can be excised, cultured, and plants regenerated. In the cell cultures of sugarcane *(Saccharum officinarum)* the phenomenon of genetic mosaic has been exploited and resistant plants to Fiji disease virus has been obtained from various selected cell lines (Krishnamurthi and Thaskal 1974). Lorz and Scowcroft (1983) demonstrated that tobacco characters induced in vitro showed Mendelian segregation, as expected for mutants.

The classical breeding method requires at least six generations for an inbred crop, and may take even longer for an outbred or vegetatively propagated crop. The development of an in vitro system to increase resistance, e.g. of a leading tetraploid potato variety, or to detect mutants or mutant clusters, had stimulated tremendous activity in unconventional breeding (Wenzel 1985).

Shepard (1981) examined the usefulness of protoplasts for creating new genotypes. He found tremendous variability in potato clones from the tetraploid Russet Burbank. Up to 30% of the regenerated protoclones were useful variants; he therefore recommended this method for intracultivar improvement. Many of the variants were grossly aberrant; some showed improved resistance to *Alternaria solani* (Matern et al. 1978) or *Phytophthora infestans* (Shepard 1981). However, not all variants were genetically stable, even during vegetative propagation; these may have originated by some change other than by genetic mutation.

Wenzel (1980) regenerated more than 3000 clones from dihaploid potatoes protoplasts. Only a few were grossly aberrant types, most of which proved to be aneuploids with no practical value (genome rather than chromosome mutations). The remainder were uniform phenotypically and in their total protein pattern. This difference in the behaviour of tetraploid and diploid potatoes shows the strong influence of the ploidy level on the survival of plants. Aneuploids are relatively more common and stable in the ordinary tetraploid than in the dihaploid potato. The degree of variation depends on the genotypes; for example Russet Burbank shows a distinctly high frequency of variants even in the field.

Somaclonal variation becomes superior to simple mutation breeding when selection pressure is applied during the in vitro phase. The common advantage that the pathogen has over the host, namely its ability to shift rapidly to a virulent population, is offset somewhat by the fact that host cells can be about as numerous as the pathogen and that flexibility for change in the host-cell population is about as great as it is in the pathogen population. However, to offset this common advantage of the pathogen requires a powerful in vitro screening system and a very large number of individual host cells (Wenzel 1985).

In general, there is no need to increase genetic variability artificially in vitro. Wenzel (1985) noted that sufficient variability already exists in most economically important plants or can be created using sexual recombination. He suggested that only to achieve certain specific traits lacking in natural sources or in specific crops, e.g. trees, with a long regeneration cycle may the in vitro approach be used. This means that, in general, tissue culture should be used to prevent rather than to increase, somaclonal variation.

g) Regeneration Within Heterogeneous Material. The most suitable source of heterogeneous cell population are microspores from a heterozygous parent. Regeneration is possible either within the anther (anther culture), Nitzsche and Wenzel (1977) or in isolated microspore culture or shed-pollen culture (Sunderland and Xu 1982).

Barley cultivars obtained by doubling haploids derived from crosses of *Hordeum bulbosum* are already on the market in Canada (Ho and G.E. Jones 1980). No variety has yet been produced using doubled androgenetic haploids; nevertheless, this approach is of increasing interest, because there are thousands of potential genotypes in the microspore population of an anther but only one in an egg cell.

Commercial barley breeders in Germany have started to use androgenetic procedures for the rapid incorporation of resistance into winter barley lines. Of particular interest is the use of haploids to incorporate resistance to soil-borne barley yellow mosaic virus (Ba YMV) into barley (Foroughi-Wehr et al. 1982, Foroughi-Wehr and Friedt 1984). More than 2000 androgenetic lines were produced from microspores from F_1 anther donor hybrids of crosses between susceptible lines and resistant cultivars such as Franka. Of 292 androgenetic lines tested, 65% were resistant to Ba YMV. This suggested that Ba YMV resistance can be efficiently and rapidly combined with other favourable traits via haploid techniques. This was done within 3 years.

This technique has been criticized for beginning with an F_1 progeny and thereby allowing only one recombination event and drastically reducing variability and the chances of new gene combinations. This is a valid criticism if the parents are genetically very different, so that several recombinations are required; however, with genetically similar varieties, which often occur, it is an invalid criticism (Wenzel 1985). In cases where the breeder intends to keep most of the genome together, the haploid procedure works. On the other hand, if frequent recombinations are necessary in order to break repulsion linkages, then the starting material should be in the F_2 or F_3 generation.

In field crops where anther culture has been successful, it has become possible to regenerate isolated microspores. Microspores are not isolated mechanically from the anther; rather, they are shed initially into liquid culture media. It has proved beneficial to obtain such pollen cultures to first condition the liquid culture medium with either anthers (Sunderland and Xu 1982) or ovules (Xu and Huang 1984) of similar genotypes. Isolated microspore cultures may offer a more effective system for regenerating a random sample from the microspore population than microspore cultured within the anther.

h) Resistant Plants Through Fusion of Protoplasts. Protoclones can be selected from protoplasts with or without prior screening systems. One of the most recent and exciting developments in the field of experimental biology, with far-reaching implications in plant improvement, is the fusion of protoplasts for obtaining somatic hybrids in incompatible crosses (Bajaj 1981). Somatic fusion allows the asexual combining of two heterozygous genomes without meiotic recombination. The hybrid plant exhibits the characteristics of both parents, depending on the degree of mixing of the cytoplasm and the nuclear fusion. This process has been termed "parasexual hybridization" (Carlson et al. 1972) because even though recombination occurs, normal gametic fusion does not take place.

The first reported event of induced protoplast fusion was by Power et al. (1970), and involved protoplasts from oat and maize. Sodium nitrate stimulated the fusion events in this study. Since that time fusion of protoplasts has been reported between several species and this topic has been reviewed by H.H. Smith (1974). A critical step in "parasexual hybridization" would seem to be regeneration of a plant from the fused cells.

The somatic fusion is important in combinations of resistances that are polygenic and inherited quantitatively. Since amphidiploid, i.e. polyploid hybrids are of no immediate practical interest, one must usually reduce the ploidy level before or after protoplast fusion to the level desired in the crop. Propagation procedures must be developed that will allow the continuous growth of a heterozygous genotype without immediate segregation (Wenzel 1985).

The problem of ploidy level may be overcome by the donor-recipient fusion technique, where one fusion partner is irradiated so as to inactivate its nuclear genome partially or completely (Aviv and Galun 1980). During fusion, only partial genomes of the two diploid parents are added to produce a diploid plant. In this way, specific parts of a foreign genome are added to a superior genome for its further improvement. In contrast to sexual recombination, the cytoplasms are also mixed and, because resistance to pathogens in several cases is coded by the genomes of plastids, fusion techniques open up the possibility for the study of cytoplasmic genetics and for producing individuals with new combinations of cytoplasmic genetic determinants (Cybrids). Incomplete fusion may also happen spontaneously, without irradiation.

i) **Disease Resistance Through Uptake of Foreign Genetic Material.** Transformation of bacteria through genetic engineering is now a routine work. This approach can be extended to isolated protoplasts of higher plants as they possess the remarkable property of taking up isolated nuclei, exogenous DNA, cytoplasmic organelles such as plastids, mitochondria and chloroplasts, and other organelles. Accordingly, the introduction of isolated DNA from disease-resistant plants into susceptible plants is worth consideration (Bajaj 1981). The first successful transfers of specific traits into higher plants being achieved, e.g. the transfer of resistance to chloramphenicol and methotrexat from bacteria into plant cells using the tumor-inducing (Ti) plasmid of *Agrobacterium radiobacter* var. *tumefaciens* as a vector (Herrera-Estrella et al. 1983) and of kanamycin resistance using cauliflower mosaic virus as the vector (Potrykus 1984). On the other hand, genetical engineering of plants resistant to pathogens is hampered by the lack of available genes, isolated or synthesized, for disease resistance (Wenzel 1985).

8 Stability and Vulnerability of Resistance

The incidence of pest and disease has been a major factor causing instability and risk in crop production. According to Swaminathan (1980), uniformity within a crop leads to genetic vulnerability, and to reinstate genetic diversity is one of the most effective means of protecting the plant against epidemics.

Individuals of a plant species dispersed among the vegetation are less infected by host-specific pathogens than when all intervening plant individuals are of the same species as themselves (Trenbath 1977). This principle applies to most of the phytophagous insects (Painter 1958; Suneson 1960). Reports are available on lower disease intensities or parasitic growth with respect to host-specific fungal parasites or susceptible plants in mixture with resistant ones (J.J. Burden and Chilvers 1976). The main reason for lighter infestation of a susceptible genotype in mixture is probably the loss of dispersing individuals or plants of the resistant (non-host) component (Leonard 1969a,b). The extra loss on non-host surfaces in the mixture (the flypaper effect) reduces the exponential population growth rate parameter, r, and thus very strongly reduces the population level attained at some given time.

Plant breeding has played an important role in increasing the yield of economically important crops. Modern crop varieties deliver high yields along with good quality. Many of the products must meet exacting industrial standards for further processing, for instance wheat for milling and baking; potatoes for frozen french fries, chips etc.; peas for freezing and canning; cotton for uniform fibre length and so on. This demand for uniformity and high yield has led to the reduction in genetic variation and with the result the older varieties in major crops, which had lower yields but considerable variation are replaced (P.R. Day 1974). Genetic homogeneity offers substantial advantages to the growers, processors, packers and consumers, which accounts for the popularity of monotypic monoculture. The main factors for farmers preferring the varieties with genetic homogeneity is an economic one, allowing them to grow one crop and variety that gives the greatest return. When there is only one best variety for a given area, everyone will grow it and this will result in genetic uniformity. Intensive agriculture also leads to uniformly dense stands so that the crop is vulnerable to any parasite. Kalyansona, a wheat variety, was released in India for cultivation in 1966. At the time of release it was resistant to yellow rust and to the prevalent races of black rust, although it was susceptible to race 122 and 42-B (complex). Since 1968, there has been a sharp rise in the distribution of race 122 and Kalyansona was reported to be susceptible in many parts of India in 1970. In 1970–71 it was reported to be susceptible to yellow rust from Lahaul Spiti valley. Extensive cultivation was one of the reasons for its becoming susceptible (Joshi et al. 1973). In the case of crops where hybrid varieties are produced from the same source of cytoplasmic male sterility, an epidemic may result if the cytoplasm becomes susceptible. A dramatic shift in the genetics of the host-parasite interaction balance occurred in the United States corn crop in 1970. In the summer of 1970, a surprise attack by southern corn leaf blight, *Drechslera maydis* brought an epiphytotic of leaf blight which has evolved from a minor disease causing an average loss of 1% to one that caused more than 12% average expected from all diseases of crops in the United States (Tatum 1971).

The factors which promote genetic vulnerability and lead to epidemics are as follows:

a) **Narrow Genetic Base.** The commercial cultivars are generally pure lines, clones or hybrid varieties, and therefore are more or less genetically uniform. The population does not have enough flexibility or buffering capacity to adapt itself to a change in the genetic make-up of the parasite, which co-evolves on it. The Irish mainly grew the susceptible variety Lumper to potato blight, and since potato varieties are vegetatively propagated, the Lumper potatoes were all alike genetically. This resulted in a total loss of the potato crop in the 1840's. Major epidemics of bacterial blight, *Xanthomonas campestris* pv. *oryzae* of rice were reported in the 1960's, after the introduction of improved high yielding varieties of rice such as Taichung Native-1, IR-8 and Jaya that did not have adequate resistance to the disease (Mew and Khush 1981).

b) **Large Acreages Under Single Cultivar.** This facilitates the spread of pathogen once infection is initiated, for instance the case of the Kalyansona variety in India as described above. Sonalika and WL 711, two cultivars of wheat, occupied large areas in India soon after their release in 1968 and 1976, respectively. Good field leaf rust resistance, a high yielding ability and bold amber grains were largely responsible for their popularity. Sonalika started showing susceptibility to leaf rust around 1975. Heavy incidence of leaf rust was observed on WL 711 in many areas around 1980. Sonalika had a recessive gene and WL 711 had a dominant gene for leaf rust resistance (A.K. Gupta et al. 1984).

c) **Introduced Parasites.** When a new parasite or a pathotype of the old parasite is introduced from another locality/geographical area(s), the local varieties may not show resistance. Introduced exotic parasites have been responsible for a number of the most devastating epidemics experienced by man. These include the late blight epidemic on potatoes in Europe in the 1840's, the chestnut blight epidemic in the United States in the early 1900's, the tropical maize rust in Africa in the 1950's and the blue mould epidemic on tobacco in the United States and Europe in the 1960's (Van der Plank 1963, Marshall 1977).

d) **Normally Innocuous Parasites to Which Marked Susceptibility has Been Introduced into Cultivars by Breeders, Along with Some Desirable Trait.** Classic examples of diseases caused by such pathogens are the *Victoria* blight epidemic on oats in the 1940's and the recent southern corn leaf blight epidemic on maize in the United States. The *Victoria* oat bred for resistance had the dominant gene for susceptibility to *Drechslera (Helminthosporium)* blight, a disease unknown up to that time.

Karnal bunt of wheat caused by *Neovossia indica* was first reported in 1931 from Northwestern India in experimental plantings. For many years it was a minor disease found only in Northwestern India. During the 1969-70 crop season it was unusually widespread in Northwest India, and since 1974-75 Karnal bunt has been distributed throughout Northern India from West Bengal to the western border. Seed lots with more than 50% of the kernels infected have been collected from farm threshing sites. The disease is now established in Afghanistan, Iraq, Pakistan and Mexico and has been intercepted in India in wheat seed that was shipped from the Lebanon and Syria. The increased distribution of Karnal bunt may be due to the development and wide distribution of wheat cultivars that were more susceptible than the older wheats that were grown in India prior to 1969-70. Resistance to Karnal bunt has been reported in several Indian wheats but most cultivars are susceptible (Joshi et al. 1983).

e) **Previously Unrecognized Strains of Pathogens/Pests Can Attack Otherwise Resistant Varieties.** The rust pathogens of cereals exist as populations of races that differ in their ability to attack various varieties. Varieties resistant when first released often become susceptible later due to the spread of previously undetected races, but the time taken for this to occur is very variable. It often occurs so rapidly as to curtail the commercial use of otherwise satisfactory varieties (R. Johnson 1978).

f) **Failure of VR.** Pathogens may overcome major gene resistance soon after a variety is released because of the intense selection pressure on the pathogen. Variants of the pathogen population that can overcome the specific resistance mechanism(s) become predominant on the variety and are considered new races of the pathogen. This can lead to a cyclic phenomenon where resistance is located, incorporated into desirable varieties, becomes widely disseminated as planting of the variety increases, a race or races of the pathogen develop that can break resistance. Epidemics occur, new sources of resistance are located and so on.

g) **Environmental Factors.** Among our problem-creating situations, there are many soil management practices that have intensified parasite control problems. An excess of nitrogen is almost uniformly deleterious to disease-escaping and disease-resistant attributes. A proper balance of nitrogen and phosphorous is a strong contributory force to escaping full impact of some of the less highly parasitic, such as the root-rot diseases. About the only situation where good crop growth and parasite resistance coincide fully is when potassium supplies approach optimum (McNew 1972).

8.1 Stability of Specific Resistance

Breeders hope that the resistant cultivars they have developed will maintain their characteristics. The results of the experiments, however, show that specific resistance, especially when controlled by a simple genetic system, is of limited value. This has been true for those genes controlling hypersensitive reaction which operate throughout the life of the plant. The resistance controlled by a single gene which operates only in the adult plant may be more lasting in its effects (I.A. Watson 1970a). The resistance of Chinese spring wheat to *Puccinia recondita* and Hope wheat to *P. graminis* could be given as examples. Any given interaction can change by gene mutation at the appropriate locus in either the host or the pathogen. Since the population of pathogens far exceeds in size that of the host, it can be assumed that for any complementary situation, detectable variability will mostly come from change in the pathogen (Watson 1970a). Flor (1958) observed that the stability of rust resistance is different from one host gene to another. In his experiments with mutant strains of flax rust, the cultivars Dakota, Cass and Polk were less stable in their resistance than Kota and Leona and Abyssinian, and these relationships held regardless of whether the variants were obtained spontaneously or by using artificial mutagens.

Borlaug (1965) remarked that the average useful life in some parts of the world of a rust-resistant variety in wheat was as short as 5 years before it succumbed to a new race of rust. Khush (1977a) reported the stability of vertical or major gene resistance in rice. Bacterial leaf blight resistant varieties with gene Xa 1 have been grown in Japan for about 30 years. In most areas the resistance has held up quite well. A strain of the bacterium virulent to Xa 1 did appear in the Beniya district of Kyushu Province (Japan) but apparently remained confined to that area. Similarly IR 20 and IR 26 having Xa 4 for resistance to blight have been grown widely for several years in the tropics and resistance has held up, except in a small area of the Philippines, where a virulent strain of bacterium to Xa4 has appeared. Bacterial blight, resistant varieties such as Benong, Sigadis, Syntha and Dewi Tara have been grown in Indonesia for 10–20 years and TKM 6, MTU 15 and Co2 in India for many years. Occurrence of bacterial blight strains virulent to these varieties under the influence of host resistance has not been reported.

Several green leafhopper (GLH) resistant varieties such as Peta, Intan, and Bengawan were widely grown in Indonesia and the Philippines for 30–35 years. Several improved plant type varieties with Glh3 gene for resistance to GLH from Peta have also been grown for several years. No clearcut evidence for the origin to GLH biotypes virulent to Glh3 has been found. The brown planthopper (BPH) resistant varieties became susceptible within 18 months of their introduction into the British Solomon Islands because of the appearance of a new biotype. Similarly, within 2 years of its large-scale cultivation in the Philippines, IR 26 was attacked by new biotypes probably originating under the influence of host resistance. The influence of host resistance on the insect populations of BPH and GLH seems to be quite different. It could be due to the differential selection pressure exerted by the resistant varieties on the insect populations.

8.2 The Proposed Solution

The alternative to uniformity is diversity. The plant breeder could work within the restrictions imposed by quality factors and within those imposed by the modes of reproduction as well, could construct varieties that satisfactorily fulfill the needs for uniformity and special needs for diversity to protect against the devastation brought by a widespread disease epidemic (M.W. Adams et al. 1971). Several measures of management of resistance gene(s) are discussed in Chap. 6, which can increase the life span of vertical gene(s). However, the breeding for multigenic resistance, synthesized HR and integrated management may reduce losses due to the disease epidemics and increase the stability of resistance are discussed in the following heads.

a) Breeding for Multi-genic Resistance. Multi-genic sources of resistance may be used because monogenic resistance is fragile due to mutation in the parasite. Single factorial resistance is the easiest to transmit and incorporate into a commerically favoured variety, but it is also the easiest to be overcome by the parasite. The multifactorial resistance may be more durable.

Resistance to *Phytophthora infestans* on potato and to *Puccinia sorghi* on maize suggests that the multi-genic non-specific resistance is more permanent. In the case of potato, multi-genic field resistance to *P. infestans* has demonstrated acceptable stability in 10 years of field trials at Toluca, where annual epiphytotics occur and many physiological races are present. In certain selections, there has been some indication of erosion of this multigenic resistance. However, no sudden breakdown has occurred, such as repeatedly reported for R-gene resistance (Niederhauser 1962). *Solanum demissum*, the species which has dominated the plant breeders, is variable in the degree of field resistance conferred on progeny. Sources of field resistance are available in several other tuber-bearing *Solanum* species. *S. stoloniferum*, *S. andigenum*, and *S. phureja* have been utilized in breeding for blight resistance in Mexico. It is reported by Niederhauser (1962) that the most practical approach to obtaining late blight resistance in commercial potato is to combine several or many of these sources of multigenic resistance in promising selections.

Pope and Dewey (1979) have suggested that quantitative resistance is universal and can be successfully manipulated. The key points suggested by them are: an assumption as a working hypothesis that (a) there is such a thing as multigenic small gene resistance; (b) this function in all conceivable and perhaps unknown biological ways would include resistance complexes that are non-functional, hence undetectable when the component genes are separated; (c) that conspicuous specific gene resistance is an integral but not necessarily essential part of these systems; and (d) the phenotypic expression of such genes considered separately has no particular relationship to their function or expression in appropriate combinations. The successful manipulation of such genes then involves doing all those things which increase the likelihood of creating and recognizing relatively rare desirable gene recombinants. This includes; (a) using genetically large populations; (b) avoiding early-generation disease screening; (c) accumulating resistant components on top of a set of partially resistant base parents, (d) adjusting disease intensity to available resistance and (e) avoiding confusion between specific and other types of disease resistance.

b) Breeding Multi-line Cultivars. The multi-line varieties may have advantages of both vertical and horizontal resistance as discussed in Chap. 6. Multi-lines may be produced and introduced in an inbreeder like oats. Their success depends on the correctness of the assumption that a super-race will not develop that may be equally pathogenic on all components of multi-line.

c) **Integrated Disease/Insect Pest Management**. The integrated management strategy may be used. This will represent suitable cultural practices, plant resistance and chemicals. However, it is essential that one should know the life cycle of the parasite, and the primary factors that influence its growth, reproduction, survival, and pathogenesis in a particular ecological setting. The complexity of developing an integrated control system could be evident by taking the example of the potato crop. Potatos are affected by more than 18 viral, 46 fungal, 6 bacterial, 5 nematode diseases and one or more parasitic seed plants, and at least 39 non-parasitic conditions. Fortunately all of them do not occur in all parts of the potato-growing regions, nor do they have the same severity in all areas. Some are foliar diseases and others are tuber and seed diseases, either in or out of the soil (Van Gundy 1972). Therefore, the integrated system must be tailored around all the diseases of a particular region and the relative importance of each.

The concept of integrated pest management was used extensively to minimize the damage and yield losses in rice in Indonesia (Dandi 1981). The resistant cultivars, primarily for brown planthopper (biotype) 1 and 2 are extensively implanted on a large scale, combined with cultural practices, such as crop rotation, simultaneous planting, sanitation by destroying the rice stubbles and lastly by the use of pesticide. Resistant varieties to the brown planthopper biotype 1 are IR 26, IR 28, IR 29, IR 30, IR 34, Ashan, Brantas, Citarum and Serayn. The resistant varieties like IR 32 and IR 38 have been introduced against biotype 2. Guidelines on the application of insecticides and rodenticides with low toxicity effect to fish and mammals recommended, effective for controlling major as well as minor pests are selected. Since the composition of the rice pest populations varies from location to location, integrated pest management practices should also be location-specific.

References

Abdel-Malek SH, Heyne EG, Painter RH (1966) Resistance of F_1 wheat plants to greenbug and hessian-fly. J Econ Entomol 59:707–710

Adams JB (1946) Aphid resistance in potatoes. Am Potato J 23:1–22

Adams MW, Ellingboe AH, Rossman EC (1971) Biological uniformity and disease epidemics. Bio Science 21:1067–1070

Afzal M, Abbas M (1943) Cotton jassid *(E. devastans)* in the Punjab V. A note on the characters of the plant associated with jassid resistance. Ind J Entomol 5:41–51

Agrawal RA, Banerjee SK, Singh M, Katiyar KN (1976) Resistance to insects in cotton. II. To pink bollworm, *Pectinophora gossypiella* (Saunders). Cotton Fibres Trop 31:217–221

Ahn SW, Ou SH (1982) Epidemiological implications of the spectrum of resistance to rice blast. Phytopathology 72:282–284

Akeson WR, Haskins FA, Gorj HJ (1969) Sweet clover weevil feeding deterrant. B. Isolation and identification. Science 163:293–294

Albersheim P, Valent B (1978) Host-pathogen interactions in plants. Plants, when exposed to oligo-saccharides of fungal origin, defend themselves by accumulating antibiosis. J Cell Biol 78:627–643

Albersio J, Lim A, Nelson MR (1975) Squash mosaic virus variability: non reciprocal cross protection between strains. Phytopathology 65:837–840

Aldwinckle HS, Lamb RC, Gustafson HL (1977) Nature and inheritance of resistance to *Gymnosporangium juniperi-virginianae* in apple cultivars. Phytopathology 67:259–266

Allan RE, Heyne EG, Jones ET, Johnston CO (1959) Genetic analysis of ten sources of hessian-fly resistance, their relationships and association with leaf rust reaction in wheat. Kans Agric Exp Stn Tech Bull 104:1–51

Allard RW (1960) Principles of plant breeding. Wiley, Toppan, New York, p 485

Allen EH, Thomas CA (1971a) A second antifungal polycetylene compound from Phytophthora-infected safflower. Phytopathology 61:1107–1109

Allen EH, Thomas CA (1971b) Trans-trans-3, 11-tridecadience-5,7,9-triyne-1, 2-diol, an antifungal polyacetylene from diseased safflower *(Carthamus tinctorius)*. Phytochemistry 10:1579–1582

Allingham EA, Jackson LF (1981) Variation in pathogenicity, virulence, and aggressiveness of *Septoria nodorum* in Florida. Phytopathology 71:1080–1085

Alston FH, Briggs JB (1968) Resistance to *Sappaphis devecta* (Wld.) in apple. Euphytica 17:468–472

Alston FH, Briggs JB (1970) Inheritance of hypersensitivity to rosy apple aphid, *Dysaphis plantaginea* in apple. Can J Genet Cytol 12:257–258

Anagnostakis SL (1971) Cytoplasmic and nuclear control of an interstrain interaction in *Ustilago maydis*. Mycologia 63:94–97

Anand SC, Gallo KM (1984) Identification of additional soybean germplasm with resistance to race 3 of the soybean cyst nematode. Plant Dis 68:593–595

Anderson AJ (1978) Isolation from three species of *Colletotrichum* of glucan-containing polysaccharides that elicit browning and phytoalexin production in bean. Phytopathology 68:189–194

Anderson EG (1956) The application of chromosomal techniques to maize improvement. Brookhaven Symp Biol 9:23–36

Anderson RG (1966) Studies on the inheritance of resistance to leaf rust of wheat. Proc Symp 2nd Int Wheat Genet Lund, Hereditas, Suppl 2:144–155

Anderson S (1961) Resistans mod havreal *Heterodera avenae*. Medd Nr 68 K Vet Landbohojk Afd Landbrugests Plantekult, Kobenhaven, 1979 (Engl Summary, pp 160–170)

Anderson-Prouty AJ, Albersheim P (1975) Host-pathogen interaction VIII. Isolation of pathogen-synthesized fraction rich glucan that elicits a defense response in the pathogen's host. Plant Physiol 56:286–291

Andrew RH, Carlson JR (1976) Preference differences of egg laying of European corn borer adults among maize genotypes (*Ostrinia nubilalis*: Lep., Pyralidae). Hortic Sci 11:143

Andrus CF (1953) Evaluation and use of disease resistance by vegetable breeders. Proc Am Soc Hortic Sci 61:434–446

Angles ER, Khush GS, Heinrichs EA (1981) New genes for resistance to white backed planthopper in rice. Crop Sci 21:47–50

Antonelli E, Daly JM (1966) Decarboxylation of indoleacetic acid by near isogenic lines of wheat resistant or susceptible to *Puccinia graminis* f.sp. *tritici*. Phytopathology 56:610–618

Apple JL (1962) Transfer of resistance to black shank (*Phytophthora parasitica* var. *nicotianae*) from *Nicotiana plumbagnifolia* to *N. tabacum*. Phytopathology 52:1 (Abstr)

Arrigoni O, Zacheo G, Arrigoni-Liso R, Bleve-Zacheo T, Lamberti F (1979) Relationship between ascorbic acid and resistance in tomato plants to *Meloidogyne incognita*. Phytopathology 69:579–581

Asada Y, Matsmumoto I (1967) Formation of lignin in the root tissues of Japanese radish affected by *Alternaria japonica*. Phytopathology 57:1339–1343

Asaga K, Higashi T (1973) A testing method for incorporating field resistance to blast with true resistance derived from foreign rice varieties. Jpn J Breed 23:152–154

Asaga K, Yoshimura S (1969) Field resistance of sister lines of rice crosses to blast disease. Ann Phytopathol Soc Jpn 35:100 (Abstr)

Athwal DS (1953) Gene interaction and the inheritance of stem rust of wheat. Ind J Genet Plant Breed 13:91–103

Athwal DS, Kimber G (1972) The pairing of an alien chromosome with homoeologous chromosome of wheat. Can J Genet Cytol 14:325–333

Athwal DS, Pathak MD (1972) Genetics of resistance to rice insects. In: Proc Symp Rice Breed IRRI, Philippines, pp 375–386

Athwal DS, Watson IW (1954) Inheritance and the genetic relationship of resistance possessed by two Kennya wheats to races of *Puccinia graminis tritici*. Proc Linn Soc NSW 79:1–4

Athwal DS, Pathak MD, Bacalangco EH, Pura CD (1971) Genetics of resistance to brown plant-hoppers and green leafhoppers in *Oryza sativa* L. Crop Sci 11:747–750

Auclair JL (1957) Developments in resistance of plants to insects. Ann Exp Entomol Soc Ontario 88:7–17

Auclair JL, Maltais JB, Cartier JJ (1957) Factors in resistance of peas to the pea aphid, *Acyrthosiphon pisum* (Harr.) (Homoptera:Aphidae). II. Amino acids. Can Entomol 89:457–464

Aujla SS, Chahal AS (1975) The screening of maize germplasms for the leaf blight of maize caused by *Helminthosporium turcicum* in Kulu Valley. Crop Improv 2(1, 2):112–117

Aujla SS, Grewal AS, Gill KS, Sharma I (1980) A screening technique for Karnal bunt disease of wheat. Crop Improv 7(2):145–146

Ausemus ER (1943) Breeding for disease resistance in wheat, oats, barley and flax. Bot Rev 9:207–260

Aviv D, Galun E (1980) Restoration of fertility in cytoplasmic (cms) *Nicotiana silvestris* fusion with X-irradiated *N. tabacum* protoplasts. Theor Appl Genet 58:121–127

Azawi AA, Campos FF (1974) Varietal resistance to some insectpests and diseases of cotton observed in central luzon. SABRAO J 6:55–59

Badwal SS (1975) Inheritance of resistance to powdery mildew in Linseed. Ind J Genet Plant Breed 35:432–433

Bagga HS, Boone DM (1968) Genes in *Venturia inaequalis* controlling pathogenicity to crobapples. Phytopathology 58:1176–1182

Bagget JR (1956) The inheritance of resistance to strains of bean yellow mosaic virus in the interspecific cross *Phaseolus vulgaris* × *P. coccineus*. Plant Dis Rep 40:702

Bahadur P, Sinha VC, Ruiker SK, Upadhyaya YM (1979) Sources of resistance to rusts and powdery mildew in wheat. Ind J Genet Plant Breed 39:402–411

Bailey JA (1973) Production of antifungal compounds in cowpea *(Vigna sinensis)* and pea *(Pisum sativum)* after virus infection. J Gen Microbiol 75:119–123

Bailey JA, Maxwell FG, Jenkins JN (1967) Boll weevil antibiosis studies with selected cotton lines utilizing egg implantation techniques. J Econ Entomol 60:1275–1279

Bailey JC (1982) Influence of plant bug and leafhopper populations on glabrous and nectariless cottons. Environ Entomol 11:1011–1013

Bajaj YPS (1981) Production of disease resistant plants through cell culture — A novel approach. J Nuclear Agric Biol 10:1–5

Bajaj YPS, Saettler AW (1970) Effect of holo toxin containing filtrates of *Pseudomonas phaseolicola* on the growth of bean callus tissue. Phytopathology 60:1065–1067

Baker EP, Teo C (1966) Mutants of *Puccinia graminis avenae* induced by ethyl methane sulphonate. Nature (London) 209:632–633

Bakhetia DRC, Bindra OS (1977) Screening technique for aphid resistance in *Brassica* crops. SABRAO J 9:91–107

Barkshale TH, Stoner AK (1977) A study of the inheritance of tomato early blight resistance. Plant Dis Rep 61:63–65

Barnes DK, Hanson CH, Frosheiser FI, Elling LR (1971) Recurrent selection for bacterial wilt resistance in alfalfa. Crop Sci 11:545–546

Barnes WC (1961) Multiple disease resistant cucumbers. Proc Am Soc Hortic Sci 77:417–423

Barnes WC, Cuthbert EB (1975) Breeding turnips for resistance to the turnip aphid. Hortic Sci 10:59–60

Barrus MF (1911) Variation of varieties of beans in their susceptibility to anthracnose. Phytopathology 1:190–195

Bartos P, Fleischmann G, Samborski DJ, Shipton WA (1969) Studies on asexual variation in the virulence of oat rust, *Puccinia coronata* f.sp. *avenae* and wheat leaf rust, *Puccinia recondita*. Can J Bot 47:1383–1387

Basavarajn R, Safeeulla KM, Murty BR (1981) Inheritance of resistance to downy mildew in pearl millet. Ind J Genet Plant Breed 41:144–149

Bassi FGA, Burnett JH (1980) The genetic architecture of aggressiveness in *Ustilago maydis*. Ann Appl Biol 94:281 (Abstr)

Batra GR, Gupta DS (1970) Screening of varieties of cotton for resistance to jassids. Cotton Grow Rev 47:285–291

Beardmore J, Ride JP, Granger JW (1983) Cellular lignification as a factor in the hypersensitive resistance of wheat to stem rust. Physiol Plant Pathol 22:209–220

Beck SD (1957) The European corn borer, *Pyrausta nubilalis* (Hubn.) and its principal host plant. VI. Host plant resistance to larval establishment. J Insect Physiol 1:158–177

Beck SD (1965) Resistance of plants to insects. Annu Rev Entomol 10:207–232

Behnke M (1979) Selection of potato callus for resistance to culture filtrates of *Phytophthora infestans* and regeneration of resistant plants. Theor Appl Genet 55:69–71

Behnke M (1980) General resistance to late blight of *Solanum tuberosum* plants regenerated from callus resistant to culture filtrates of *Phytophthora infestans*. Theor Appl Genet 56:151–152

Benepol PS, Hall CV (1967) The genetic basis of varietal resistance of *Cucurbita pepo* L. to squash bug *Anasa tristis* (De Geer). Proc Am Soc Hortic Sci 90:301–303

Bhojwani SS, Evans PK, Cocking EC (1977) Protoplast technology in relation to crop plants: Progress and problems. Euphytica 26:343–360

Biffen RH (1905) Mendel's laws of inheritance and wheat breeding. J Agric Sci 1:4–48

Biffen RH (1907) Studies on inheritance of disease resistance. J Agric Sci 2:109–128

Biffen RH (1912) Studies on inheritance of disease resistance II. J Agric Sci 4:421–429

Black JH, Leigh TF (1963) The biology of the boll weevil in relation to cotton types. J Econ Entomol 56:789–790

Black W (1960) Races of *Phytophthora infestans* and resistance problems in potatoes. Scot Plant Breed Stn Annu Rep 1960:29–38

Blanch PA (1980) Inheritance of pathogenicity in *Gaeumannomyces graminis* var. *tritici*. Ann Appl Biol 94:301 (Abstr)

Blum A (1972) Sorghum breeding for shoot fly resistance in Israel. In: Jotwani MG, Young WR (eds) Control of sorghum shoot fly. Oxford, pp 180–191

Bohn GW, Tucker CM (1939) Studies on *Fusarium* wilt of tomato. I. Immunity in *Lycopersicon pimpinellifolium* Mill. and its inheritance in hybrids. Science 89:603

Bohn GW, Kishaba AN, Principe JA, Toba HH (1973) Tolerance to melon aphid in *Cucumis melo.* J Am Soc Hortic Sci 98:37–40

Bolley LH (1905) Breeding for resistance or immunity to disease. Proc Am Breed Assoc 1:131–135

Bonde MR, Miller RL, Ingham JL (1973) Induction and identification of sativan and vesitol as two phytoalexins from *Lotus corniculatus*. Phytochemistry 12:2957–2959

Boone DM, Keitt GW (1957) *Venturia inaequalis* (Cke.) Wint. XII. Genes controlling pathogenicity of wild-type lines. Phytopathology 47:403–409

Boozaya-Angoon D, Starks KJ, Edwards LH, Pass H (1981) Inheritance of resistance in oats to two biotypes of the green bug. Environ Entomol 10:557–559

Borikar ST, Chopde PR (1980) Inheritance of shoot fly resistance under three levels of infestation in sorghum. Maydica XXV:175–183

Borlaug NE (1959) The use of multilineal or composite varieties to control air borne epidemic diseases of self-pollinated crop plants. In: Proc Symp 1st Int Wheat Genet, Winnipeg, pp 12–27

Borlaug NE (1965) Wheat, rust, and people. Phytophatology 55:1088–1098

Borlaug NE (1983) Contributions of conventional plant breeding to food production. Science 219: 689–693

Borlaug NE, Gibler JW (1953) The use of flexible composite wheat varieties to control the constantly changing stem rust pathogen. Agron Abstr p 81

Bos L (1980) Introduction to plant virology. Oxford and IBH, Oxford, p 160

Bottino PJ (1975) The potential of genetic manipulation in plant cell cultures for plant breeding. Radiat Bot 15:1–16

Boukema IW (1982) Inheritance of resistance to *Didymella lycopersici* Klebb. in tomato (*Lycopersicon* Mill). Euphytica 31:981–989

Boukema IW (1983) Inheritance of resistance to foot and root rot caused by *Phytophthora nicotianae* v. Breda De Hann var. *Nicotianae* in tomato (*Lycopersicon* Mill.). Euphytica 32:103–109

Bozzini A (1971) First results of bunt resistance analysis in mutants of durum wheat. In: Proc Symp Mutat Breed Dis Resist. Int At Energy Ag, Vienna, pp 131–137

Bracker CE, Littlefield LJ (1973) Structural concept of host-pathogen interfaces. In: Byrde RJW, Cutting CV (eds) Fungal pathogenicity and the plants response. Academic Press, London New York, pp 153–313

Brazzel JR, Martin DF (1957) Oviposition sites of the pink bollworm on the cotton plant. J Econ Entomol 50:122–124

Briggle LW, Vogel OA (1968) Breeding short stature disease resistant wheats in United States. Euphytica (Suppl) 1:107–130

Briggs JB (1959) Three new strains of *Amphorophora rubi* (Kalt.) on cultivated raspberries in England. Bull Entomol Res 50:81–87

Briggs JB (1965) The distribution, abundance, and genetic relationships of four strains of the rubis aphid (*Amphorophora rubi* Kalt.) in relation to raspberry breeding. J Hortic Sci 40:109–117

Briggs FN, Allard RW (1953) The current status of the backcross method of plant breeding. Agron J 45:131–138

Briggs SP, Scheffer RP, Haug AH (1984) Selective toxin of *Helminthosporium victoriae*: Thermal relationships in effects on oat tissues and protoplasts. Phytopathology 74:768–773

Brinkerhoff LA (1970) Variation in *Xanthomonas malvacearum* and its relation to control. Annu Rev Phytopathol 8:85–110

Bromfield KR (1961) The effect of post inoculation temperature on seedling reaction of selected wheat varieties to stem rust. Phytopathology 51:590–593

Bromfield KR, Hartwig EE (1980) Resistance to soybean rust and mode of inheritance. Crop Sci 20:254–255

Browder LE, Lyon FL, Eversmeyer MG (1980) Races, pathogenicity, phenotypes, and type cultures of plant pathogens. Phytopathology 70:581–583

Browning JA, Frey KJ (1959) The inheritance of new sources of stem rust resistance. Plant Dis Rep 43:768–771

Browning JA, Frey KJ (1969) Multiline cultivars as a means of disease control. Annu Rev Phyto-pathology 7:355–382

Browning JA, Simons MD, Frey KJ (1962) The potential value of synthetic tolerant or multiline varieties for control of cereal rusts in north America. Phytopathology 52:726 (Abstr)

Browning JA, Simons MD, Grindeland RL (1964) Breeding multiline oat varieties from Iowa. Iowa Farm Sci 19(8):5–8

Bukasov SM (1936) The problems of potato breeding. Am Potato J 13:235–252

Burden JJ, Chilvers GA (1976) Epidemiology of *Pythium* induced damping off in mixed species seedling stands. Ann Appl Biol 82:233–240

Burden RS, Bailey JA (1975) A structure of the phytoalexin from soybean. Phytochemistry 14:1389–1390

Burden RS, Bailey JA, Dawson GW (1972) Structures of three new isoflavonoids from *Phaseolus vulgaris* infected with tobacco necrosis virus. Tetrahedron Lett 1972:4175–4178

Burk LG (1967) An interspecific bridge cross, *Nicotiana repanda* through *N. sylvestris* to *N. tabacum*. J Hered 58:215–218

Buzzell RI, Tu JC (1984) Inheritance of soybean resistance to soybean mosaic virus. J Hered 75:82

Byther RS, Steiner GW (1972) Use of helminthosporoside to select sugarcane seedlings resistant to eye spot disease. Phytopathology 62:466–470

Caldwell RM (1966) Advances and challenges in the control of plant diseases through breeding. In: Proc Symp Pest Control Chem, Biol, Genet Phys Means. AAAS US Dep Agric ARS, Washington, pp 33–110

Caldwell RM (1968) Breeding for general and/or specific plant disease resistance. In: Proc Symp 3rd Int Wheat Genet, Canberra, pp 263–272

Caldwell RM, Cartwright WB, Compton LE (1946) Inheritance of hessian-fly resistance derived from W 38 and durum PI.94587. Am Soc Agron J 38:398–409

Caldwell RM, Schafer JE, Compton LE, Patterson FL (1957) A mature plant type of wheat leaf-rust resistance of composite origin. Phytopathology 47:691–692

Caldwell RM, Schafer JE, Compton LE, Patterson FL (1958) Tolerance to cereal leaf rusts. Science 128:714–715

Caldwell RM, Cartwright WB, Compton LE (1959) Dual, a hessian-fly resistant soft and winter wheat. Purdue Univ Agric Exp Stn (Mimeo) ID-44:1

Caldwell BE, Brim CA, Ross JP (1960) Inheritance of resistance of soybeans to the cyst nematode *Heterodera glycines*. Agron J 52:635–636

Caldwell RM, Gallun RL, Compton LE (1966) Genetics and expression of resistance to hessian-fly, *Phytophaga destructor* (Say). In: Proc Symp 2nd Int Wheat Genet, Lund, 1963. Hereditas (Suppl) 2:462–463

Campbell TA (1984) Inheritance of seedling resistance to gray mold in Kenaf. Crop Sci 24:733–734

Cantello WW, Boswell AL, Argauer RJ (1974) Tetranychus mite repellent in tomato. Environ Ento-mol 3:128–130

Carlson PS (1973) The use of protoplasts for genetic research. Proc Natl Acad Sci USA 70:598–602

Carlson PS, Smith HH, Dearing RD (1972) Parasexual interspecific plant hybridization. Proc Natl Acad Sci USA 69:2292–2294

Carnahan HL, Peaden RN, Lieberman FV, Pettersen RK (1963) Differential reactions of alfalfa varieties and selections to the pea aphid. Crop Sci 3:219–222

Carroll RB, Lukejic FL, Levine RG (1972) Absence of a common antigen relationship between *Corynebacterium insidiosum* and *Medicago sativa* as a factor in disease development. Phyto-pathology 62:1351–1360

Carson ML, Hooker AL (1981) Inheritance of resistance to stalk rot of corn caused by *Colletotrichum graminicola*. Phytopathology 71:1190–1196

Cartier JJ (1963) Varietal resistance of peas to pea aphid genotypes under field and glasshouse con-ditions. J Econ Entomol 56:205–213

Cartier JJ, Isaak A, Painter RH, Sorensen EL (1965) Biotypes of pea aphid, *Acyrthosiphon pisum* (Harris) in relation to alfalfa clones. Can Entomol 97:754–760

Cartwright WB, Weibe GA (1936) Inheritance of resistance to the hessian-fly in the wheat crosses Dawson × Poso and Dawson × Big Club. J Agric Res 52:691–695

Casagrande RA, Haynes DL (1976) The impact of pubescent wheat on the population dynamics of the cereal beetle. Environ Entomol 5:153–159

Chahal SS, Gill KS, Phul PS, Aujla SS (1975) Screening of the pearl millet germplasm for the green ear disease, caused by *Sclerospora graminicola* Sacc. Crop Improv 2(1,2):65–70

Chahal SS, Gill KS, Phul PS, Singh NB (1981) Effectiveness of recurrent selection for generating ergot resistance in pearl millet. SABRAO J 13:184–186

Chalfant RB, Gaines TP (1973) Cowpea curculio: Correlations between chemical composition of the southern pea and varietal resistance. J Econ Entomol 66:1011–1013

Chambliss OL, Jones CM (1966a) Chemical and genetic basis for insect resistance in cucurbits. Proc Am Soc Hortic Sci 89:394–405

Chambliss OL, Jones CM (1966b) Cucurbitacins: Specific insect attractants in cucurbitaceae. Science 153:1392–1393

Chang TT, Ou S, Pathak MD, Ling KC, Kauffman HE (1975) The search for disease and insect resistance in rice germplasm. In: Frankel OH, Hawkes JG (eds) Crop genetic resources for today and tomorrow. Cambridge Univ Press, Cambridge, pp 183

Chang WL (1975) Inheritance of resistance to brown planthoppers in rice. SABRAO J 7:53–60

Chang WL, Chen LC (1971) Resistance of rice varieties to the brown plant hoppers (*Nilaparvata lugens* Stal.). J Taiwan Agric Res 20:12–20

Chaudhary RRP, Bhattacharya AK (1982) Natural resistance of soybean flour to *Trogoderma granarium* (Everts.). Bull Grain Tech XX:175–178

Chen LC, Chang WL (1971) Inheritance of resistance to brown plant hoppers in rice variety Mudgo. J Taiwan Agric Res 20:57–60

Cheng CH, Pathak MD (1972) Resistance to *Nephotettix virescens* in rice varieties. J Econ Entomol 65:1148–1153

Chiang HS, Taleker NS (1980) Identification of sources of resistance to the beanfly and two other agromyzid flies in soybean and mungbean. J Econ Entomol 73:197–199

Chiang MS, Crete R (1976) Diallel analysis of the inheritance to race 6 of *Plasmodiophora brassicae* in cabbage. Can J Plant Sci 56:865–868

Chiang MS, Crete R (1983) Transfer of resistance to race 2 of *Plasmodiophora brassicae* from *Brassica napus* to cabbage (*B. oleracea* ssp. *capitata*). V. The inheritance of resistance. Euphytica 32:479–483

Chiang MS, Hudson M (1973) Inheritance of resistance to the European corn borer in grain corn. Can J Plant Sci 53:779–782

Clayton EE (1947a) New kinds of tobacco. In: US Dep Agric. Yearb Agric, Washington, pp 363

Clayton EE (1947b) A wildfire resistant tobacco. J Hered 38:35–40

Clayton EE (1958) The genetics and breeding progress in tobacco during the last 50 years. Agron J 50:352–356

Cockerham G (1943) Potato breeding for virus resistance. Ann Appl Biol 30:105–108

Cocking EC (1972) Plant cell protoplasts-isolation and development. Annu Rev Plant Physiol 23:29–50

Codon P, Kuc J (1962) Confirmation of the identity of a fungi toxic compound produced by carrot root tissue. Phytopathology 52:182–183

Cody CE (1941) Colour preference of the pea aphid in western Oregon. J Econ Entomol 34:584

Comstock JC, Scheffer RP (1973) Role of host-selective toxin in colonization of corn leaves by *Helminthosporium carbonum*. Phytopathology 63:24–29

Cook OF (1906) Weevil resisting adaptations of cotton plant. US Dep Agric Bull 88

Coons GH (1954) The wild species of *Beta*. Proc Am Soc Sugar Beet Technol 8(2):142

Costa CP de, Jones CM (1971) Cucumber beetle resistance and mite susceptibility controlled by the bittergene in *Cucumis sativus* L. Science 172:1145–1146

Cotter RU, Roberts BJ (1963) A synthetic hybrid of two varieties of *Puccinia graminis*. Phytopathology 53:344–346

Cournoyer BM (1970) Crown rust epiphytology with emphasis on the quantity and periodicity of spore dispersed from heterogeneous oat cultivare-rust race populations, dissertation. Iowa State Univ Lib Ames

Coyne DP, Schuster ML (1974) Breeding and genetics of tolerance of several bean (*Phaseolus vulgaris* L.) bacterial pathogens. Euphytica 23:651–656

Craig J (1976) An inoculation technique for identifying resistance to sorghum downy mildew. Plant Dis Rep 60:350–352

Crill JP, Burgis DS, Strobel JW (1971) Development of multiple disease-resistant fresh market tomato varieties adapted for machine harvest. Hortic Sci 61:888

Crill P (1977) An assessment of stabilizing selection in crop variety development. Annu Rev Phytopathol 15:185–202

Cruickshank IAM, Perrin D (1960) Isolation of a phytoalexin from *Pisum sativum* L. Nature (London) 187:799–800

Cruickshank IAM, Perrin DR (1965) Studies on phytoalexins. IX. Pisatin formation by cultures of *Pisum sativum* L. and several other *Pisum* species. Aust J Biol Sci 18:829–835

Cruickshank IAM, Perrin DR (1971) Studies on phytoalexins. XI. The induction, antimicrobial spectrum and chemical assay of phaseollin. Phytopathol Z 70:209–229

Cruickshank IAM, Veeraraghavan J, Perrin DR (1974) Some physical factors affecting the formation and/or net accumulation of medicarpin in infection droplets on white clover leaflets. Aust J Plant Physiol 1:149–156

Curtis BC, Schlenhuber AM, Wood EA Jr (1960) Genetics of green bug (*Toxoptera graminum* R.) resistance in two strains of common wheat. Agron J 52:599–602

Cuthbert FP Jr, Davis BW (1972) Factors contributing to cowpea curculio resistance in southern peas. J Econ Entomol 65:778–781

Dabholkar AR, Baghel SS (1980) Inheritance of resistance to grain mould of sorghum. Ind J Genet Plant Breed 40:472–475

Dahlbeck D, Stahl RE (1979) Mutation for change of race in culture of *Xanthomonas vesicatoria*. Phytopathology 69:634–636

Dahlbeck D, Stahl RE, Jones JP (1979) The effect of vertical and horizontal resistance on development of bacterial spot of pepper. Plant Dis Rep 63:332–335

Dale MFB, Philips MS (1982) An investigation of resistance to the white potato cyst-nematode. J Agric Sci 99:325–328

Dandi S (1981) Implementation of integrated pest management for rice in Indonesia. Abstr Int Conf Trop Crop Prot, Lyon, France 8–10 July, 1981, p 52

Daniels NE, Porter KB (1958) Green bug resistance studies in winter wheat. J Econ Entomol 51:702–704

Darwin C (1868) The variation of animals and plants under domestication, vol I. Murray, London, p 354

Das YT (1976) Cross resistance to stem borers in rice varieties. J Econ Entomol 69:41–46

Dash AN, Dikshit UN (1982) Sources of resistance in jute germplasm against yellow mite. Ind J Genet Plant Breed 42:87–91

Daubeny HA (1966) Inheritance of immunity in the red raspberry to the north American strain of the aphid, *Amphorophora rubi* (Kalt.) Proc Am Soc Hortic Sci 88:346–351

Daubeny HA (1972) Breeding for aphid immunity in the red raspberry. Hortic Sci 7:327 (Abstr)

Day PK (1960) Variation in phytopathogenic fungi. Annu Rev Microbiol 14:1–16

Day PR (1956) Race names of *Cladosporium fulvum*. Tomato Genet Crop Rep 6:13–14

Day PR (1957) Mutation to virulence in *Cladosporium fulvum*. Nature (London) 179:1141–1142

Day PR (1974) Genetics of host-parasite interaction. Freeman, San Francisco, p 238

Dayani TR, Bakshi JS (1978a) Inheritance of resistance to stripe rust in barley. Ind J Genet Plant Breed 38:283–284

Dayani TR, Bakshi JS (1978b) Inheritance of resistance to corn leaf aphid in barley. Ind J Genet Plant Breed 38:281–282

Deaton WR, Keyes GJ, Collins GB (1984) Expressed resistance to black shank among tobacco callus cultures. Theor Appl Genet 63:65–70

Dessert JM, Baker LR, Fobes JF (1982) Inheritance of reaction to *Pseudomonas lachrymans* in pickling cucumber. Euphytica 31:847–855

Deverall BJ (1976) Current perspectives in research on phytoalexins. In: Friend J, Threlfall DR (eds) Biochemical aspects of plant parasite relationships. Academic Press, London New York, pp 207–223

Deverall BJ (1977) Defence mechanisms of plants. Cambridge Univ Press, Cambridge, p 110

Devine TE, Hanson CH (1973) Hardy and resistant alfalfa. Agric Res 21(10):11

Diachun S, Hensen L (1959) Inheritance of necrotic, mottle, and resistant reaction of bean yellow mosaic virus in clones of red clover. Phytopathology 49:537 (Abstr)

Dickson MH, Boettger MA (1977) Breeding for multiple root rot resistance in snapbeans. J Am Soc Hortic Sci 102:373–376

Dickson MK, Eckenrode CJ (1975) Variation in *Brassica oleracea* resistance to cabbage looper and imported cabbage worm in the green house field. J Econ Entomol 68:757–776

Dijkman A van, Sijpesteijn AK (1971) A biochemical mechanism for the gene-for-gene resistance of tomato to *Cladosporium fulvum*. Neth J Plant Pathol 77:14–24

Djamin A, Pathak MD (1968) Role of silica in resistance to Asiatic borer, *Chilo suppressalis* (Walker), in the rice varieties. J Econ Entomol 60:347–351

Dobias K (1978) Resistance of some potato species to black leg. Potato Abstr 1981:557

Doubly JA, Flor HH, Clagget CO (1960) Relation of antigens *Melampsora lini* and *Linum usitatissimum* to resistance and susceptibility. Science 131:229

Dougall DK (1972) Cultivation of plant cells. In: Rothblat GH, Cristofalo VJ (eds) Growth, nutrition and metabolism of cells in culture. Academic Press, London New York, pp 371–406

Douglas WA, Eckhardt RC (1957) Dent corn inbreds and hybrids resistant to the corn ear worm in the south. US Dep Agric Tech Bull 1160:16

Dovas C, Skylakakis G, Georgopulos SG (1976) The adaptability of the benomyl-resistant population of *Cercospora beticola* in Northern Greece. Phytopathology 66:1452–1456

Drijfhout E (1978) Genetic interaction between *Phaseolus vulgaris* and bean common mosaic virus with implications for strain identification and breeding for resistance. Versl Landbouwkd Onderz Agric Res Rep 872:1–98

Driscoll CJ, Jensen NF (1963) A genetic method for detecting induced intergeneric translocations. Genetics 48:459–468

Driscoll CJ, Jensen NF (1964) Characteristics of leaf rust resistance transferred from Rye to Wheat. Crop Sci 4:373–374

Dudley JW, Hill RR Jr, Hanson CH (1963) Effect of seven cycles of recurrent phenotypic selection on means and genetic variances of several characters in two pools of alfalfa germplasm. Crop Sci 3:543–546

Dunn GM, Namm T (1970) Gene dosage effects on monogenic resistance to northern corn leaf blight. Crop Sci 10:352–354

Dunn JA (1960) Varietal resistance of lettuce to attack by the lettuce root aphid, *Pemphigus bursarius* (L.). Ann Appl Biol 48:764–770

Dunn JA (1974) Study on inheritance of resistance to root aphid, *Pemphigus bursarius*, in lettuce. Ann Appl Biol 76:9–18

Dunn JA, Kempton DPH (1976) Varietal differences in the susceptibility of brussels sprouts to Lepidopterous pests. Ann Appl Biol 82:11–19

Dunnett JM (1957) Variation in pathogenicity of the potato root eelworm (*Heterodera rostochiensis* Woll.) and its significance in potato breeding. Euphytica 6:77–89

Dunnett JM (1960) Potato breeders strains of root eelworm (*Heterodera rostochiensis* Woll.). Rep 5th Int Plant Nematol 10–13 Aug. 1959 Uppsala. Nematologia (Suppl):84–94

Dutrecq AJE (1977) In-vitro selection of plants resistant or tolerant to pathogenic fungi. In: Proc Symp Induced Mutat Against Plant Dis. Int At Energy Ag, Vienna, pp 471–478

Dutt KVLN, Seshu DV, Shastry SVS (1980) Inheritance of resistance to stem borer in rice. Ind J Genet Plant Breed 40:166–171

Dvorak J (1977) Transfer of leaf rust resistance from *Aegilops speltoides* to *Triticum aestivum*. Can J Genet Cytol 19:133–141

Dwivedi S, Singh DP (1985) Inheritance of resistance to yellow mosaic virus in a wide cross of blackgram (*Vigna mungo* (L.) Hepper). Z Pflanzenzücht 95:281–284

Dyck PL, Kerber ER (1970) Inheritance in hexaploid wheat of adult-plant leaf rust resistance derived from *Aegilops squarrosa*. Can J Genet Cytol 12:175–180

Dyck PL, Samborski DJ (1982) The inheritance of resistance to *Puccinia recondita* in a group of common wheat cultivars. Can J Genet Cytol 24:273–283

Dyck PL, Samborski DJ, Anderson RG (1966) Inheritance of adult-plant leaf rust-resistance derived from the common wheat varieties. Exchange and rontana. Can J Genet Cytol 8:665–671

Eamchit S, Mew TW (1982) Comparison of virulence of *Xanthomonas campestris* pv. *oryzae* in Thailand and the Philippines. Plant Dis 66:556–559

Eckroth EG, McNeal FH (1953) Association of plant characters in spring wheat with resistance to the wheat stem sawfly. Agron J 45:400–404

Eenink AH (1977) Genetics of host-parasite relationship and the stability of resistance. In: Proc Symp Induced Mutat Against Plant Dis. Int At Energy Ag, Vienna, pp 47–57

Eenink AH, Dieleman FL, Groenwold R (1982a) Resistance of lettuce *(Lactuca)* to the leaf aphid, *Nasonovia ribis nigri* 2. Inheritance of the resistance. Euphytica 31:301–304

Eenink AH, Groenwold R, Dieleman FL (1982b) Resistance of lettuce *(Lactuca)* to the leaf aphid, *Nasonovia ribis nigri* 1. Transfer of resistance from *L. virosa* to *L. sativa* by interspecific crosses and selection of resistant breeding lines. Euphytica 31:291–300

Elgersma DM (1980) Accumulation of rishitin in susceptible and resistant tomato plants after inoculation with *Verticillium albo-atrum*. Physiol Plant Pathol 16:149–153

Ellenby C (1954) Tuber forming species and varieties of the genus *Solanum* tested for resistance to the potato-root eelworm, *Heterodera rostochiensis* Wollemweber. Euphytica 3:195–202

Elliot FC (1967) X-ray induced translocation of *Agropyron* stem rust resistance to common wheat. J Hered 48:77–87

Elliot JEM, Whittington WJ (1979) An assessment of varietal resistance to chocolate spot *(Botrytis fabae)*, infection of field beans (*Vicia faba* L.), with some indications of its heritability and mode of inheritance. J Agric Sci 93:411–417

Emden HF van (1966) Plant insect relationship and pest control. World Rev Pest Control 5:115–123

Erickson J (1894) Über die Spezialisierung des Parasitismus bei den Getreiderostpilzen. Ber Deut Bot Ges 12:292

Erickson JM, Fenny P (1974) Sinigrin: A chemical barrier to the black swallowtail butterfly, *Papilio polyxenes*. Ecology 55:103–111

Etten HD van, Pueppke SG (1976) Isoflavonoid phytoalexins. In: Friend J, Threlfalls DR (eds) Biochemical aspects of plant-parasite relationships. Academic Press, London New York, pp 239–289

Everly RT (1967) Establishment and development of corn leaf aphid populations on inbred and single cross dent corn. Proc North Cent Brauch Entomol Soc Am 22:80–84

Faluyi JO, Olorode O (1984) Inheritance of resistance to *Helminthosporium maydis* blight in maize *(Zea mays)*. Theor Appl Genet 67:341–344

Fan Z, Rimmer SR, Stefansson BR (1983) Inheritance of resistance to *Albugo candida* in rape (*Brassica napus* L.). Can J Genet Cytol 25:420–424

Farrar W (1898) The making and improvement of wheats for Australian conditions. Agric Gaz NS Wales 9:131–168

Fassuliotus G (1970) Resistance of *Cucumis* species to root knot nematode, *Meloidogyne incognita acrita*. J Nematol 2:174–178

Fatunla T, Badaru K (1983) Inheritance of resistance to cowpea weevil (*Callosobruchus maculatus*, Fabr.). J Agric Sci 101:423–426

Favret EA (1971a) Different categories of mutations for disease reaction in the host organism. In: Proc Symp Mutat Breed Dis Resist Int At Energy Ag, Vienna, pp 107–116

Favret EA (1971b) Basic concepts on induced mutagenesis for disease reaction. In: Proc Symp Mutat Breed Dis Resis. Int At Energy Ag, Vienna, pp 55–65

Fawcett CH, Firn RD, Spencer DM (1971) Wyerone increase in leaves of broad bean (*Vicia faba* L.) after infection by *Botrytis fabae*. Physiol Plant Pathol 1:163–166

Fehrmann H, Dimond AE (1967) Peroxidase activity and *Phytophthora* resistance in different organs of the potato plant. Phytopathology 57:69–72

Fenny P (1970) Seasonal changes in oak leaf tannins and nutrients as a cause of spring feeding by winter moth caterpillars. Ecology 51:565–581

Ferguson S, Sorensen EL, Horber EK (1982) Resistance to the spotted alfalfa aphid (Homoptera: Aphididae) in glandular haired *Medicago* species. Environ Entomol 11:1229–1232

Ferris VR (1955) Histological study of pathogen-suscept relationships between *Phytophthora infestans* and derivatives of *Solanum demissum*. Phytopathology 45:546–552

Fery RL, Dukes PD, Cuthbert FP Jr (1976) The inheritance of *Cercospora cruenta* leaf spot resistance in the southern pea (*Vigna unguiculata* (L.) Walp.). J Am Soc Hortic Sci 101:148–149

Fick GN, Zimmer DE (1974) Monogenic resistance to *Verticillium* wilt in sunflowers. Crop Sci 14:895–896

Fiddian WEH, Kimber DS (1964) A study of biotypes of cereal cyst nematode (*Heterodera avenae* Woll.) in England and Wales. Nematologia 10:631–636

Figoni RA, Rutger JN, Webster RK (1983) Evaluation of wild *Oryza* species for stem rot (*Sclerotium oryzae*) resistance. Plant Dis 67:998–1000

Fincham JRS, Day PR (1971) Fungal genetics, 3rd edn. Blackwells, Oxford

Finlay KW (1953) Inheritance of spotted wilt resistance in tomato. Five genes controlling spotted wilt resistance in four tomato types. Aust J Biol Sci 6:153

Flangas AL, Dickson JG (1957) A method for detailed genetic analysis of pathogenic loci in rust fungi, and genetic analysis of pathogenicity in *Puccinia sorghi* (Abstr.). Phytopathology 47:521

Flor HH (1942) Inheritance of pathogenicity in *Melampsora lini*. Phytopathology 32:653–669

Flor HH (1947) Inheritance of reaction to rust in flax. J Agric Res 74:241–262.

Flor HH (1955) Host-parasite interaction in flax rust-its genetics and other implications. Phytopathology 45:680–685

Flor HH (1956a) The complementary genetic system in flax rust. Adv Genet 8:29–59

Flor HH (1956b) Mutations in flax rust induced by ultraviolet radiation. Science 124:888–889

Flor HH (1958) Mutation to wider virulence in *Melampsora lini*. Phytopathology 48:297–301

Flor HH (1959) Differential host range of the monocaryon and dicaryons of a eu-autoecius rust. Phytopathology 49:794–795

Flor HH (1960a) A sexual variants of *Melampsora lini*. Phytopathology 50:223–226

Flor HH (1960b) The inheritance of X-ray induced mutations to virulence in a urediospore culture of race 1 of *Melampsora lini*. Phytopathology 50:603–605

Flor HH (1971) Current status of gene for gene concept. Annu Rev Phytopathol 9:275–296

Flor HH, Comstock VE (1971) Development of flax cultivars with multiple-rust conditioning genes. Crop Sci 11:64–66

Florica N (1971) Subdivision of races of *Puccinia recondita* f.sp. *tritici* distributed in Rumania during 1964–67. An Inst Cerel Prot Plant 7:41–48

Foroughi-Wehr B, Friedt W (1984) Rapid production of recombinant barley yellow mosaic virus resistant *Hordeum vulgare* lines by anther culture. Theor Appl Genet 67:377–382

Foroughi-Wehr B, Friedt W, Wenzel G (1982) On the genetic improvement of androgenetic haploid formation in *Hordeum vulgare*. Theor Appl Genet 62:233–239

Fraile A, Garcia-Arenal F, Garcia-Serrano JJ, Sagasta EM (1982) Toxicity of phaseollin, phaseollidin, phaseollinisoflavan and kievitone to *Botrytis cinerea*. Phytopathol Z 105:161–169

Frandsen NO (1956) Rasse 4 von *Phytophthora infestans* in Deutschland. Phytopathol Z 26:124–130

Frank JA, Paxton JD (1971) An inducer of soybean phytoalexin and its role in the resistance of soybeans to *Phytophthora* rot. Phytopathology 61:954–958

Frankel OH (1977) Genetic resources as the backbone of plant protection. In: Proc Symp Induced Mutat Against Plant Dis. Int At Energy Ag, Vienna, pp 3–14

Franzone PM, Favret EA (1982) Mutations for pathogenicity in wheat leaf rust *Puccinia recondita* var. *tritici* induced by ethyl-methane sulphonate. Phytopathol Z 104:289–298

Frey KJ, Browning JA (1971) Breeding crop plants for disease resistance. In: Proc Symp Mutat Breed Dis Resist. Int At Energy Ag, Vienna, pp 45–54

Frey KJ, Browning JA, Grindeland RL (1971) Implementation of oat multiline cultivar breeding. In: Proc Symp Mutat Breed Dis Resist. Int At Energy Ag, Vienna, pp 159–169

Frey KJ, Browning JA, Simons MD (1973) Management of host resistance genes to control disease. Z Pflanzenkrankh Pflanzensch 80:160–180

Frey KJ, Browning JA, Simons MD (1975) Multiline cultivars of autogamous crop plants. SABRAO J 7:113–123

Frey KJ, Browning JA, Simons MD (1979) Management systems for host genes to control disease loss. Ind J Genet Plant Breed 39:10–21

Friend WG (1958) Nutritional requirements of phytophagous insects. Annu Rev Entomol 3:57–74

Fullar JM, Howard HW (1974) Breeding for resistance to white potato cyst nematode, *Heterodera pallida*. Ann Appl Biol 77:121–128

Fuller PA, Coyne DP, Steadman JR (1984) Inheritance of resistance to white mold disease in a diallel cross of dry beans. Crop Sci 24:929–933

Futrell MC, Hooker AL, Scott GE (1975) Resistance in maize to corn rust, controlled by a single dominant gene. Crop Sci 15:597–599

Gallun RL (1965) The hessian-fly. US Dep Agric Leafl 533:8

Gallun RL (1972) Genetic interrelationships between host plants and insects. J Environ Qual 1:259–265

Gallun RL (1977) Genetic basis of hessian-fly epidemics. Ann NY Acad Sci 287:223–229

Gallun RL (1978) Genetics of biotypes B and C of the hessian-fly. Ann Entomol Soc Am 71:481–486

Gallun RL, Hatchett JH (1968) Interrelationship between races of hessian-fly, *Mayetiola destructor* (Say) and resistance in wheat. In: Proc Symp 3rd Int Wheat Genet. Aust Acad Sci, Canberra, pp 258–262

Gallun RL, Hatchett JH (1969) Genetic advance of elimination of chromosomes in the hessian-fly. Ann Entomol Soc Am 62:1095–1101

Gallun RL, Khush GS (1979) Genetic factors affecting expression and stability of resistance. In: Maxwell FG, Jennings PR (eds) Breeding plants resistance to insects. Wiley Interscience, New York, pp 64–85

Gallun RL, Patterson FL (1977) Monosomic analysis of wheat for resistance to hessian-fly. J Hered 68:223–226

Gallun RL, Reitz LP (1971) Wheat cultivars resistant to races of hessian-fly. US Dep Agric Prod Res Rep 134:16

Gallun RL, Ruppel R, Everson EH (1966) Resistance of small grains to the cereal leaf beetle. J Econ Entomol 59:827–829

Gallun RL, Roberts JJ, Finney RE, Patterson FL (1973) Leaf pubescence of field grown wheat: a deterrent to oviposition by cereal leaf beetle. J Environ Qual 2:333–334

Gardenhire JH (1964) Inheritance of resistance in oats. Crop Sci 4:443

Gardenhire JH (1965) Inheritance and linkage studies on green bug resistance in barley (*Hordeum vulgare* L.). Crop Sci 5:28–29

Gardenhire JH, Chada HL (1961) Inheritance of green bug resistance in barley. Crop Sci 1:349–359

Gardenhire JH, Tuleen NA, Stewart KW (1973) Trisomic analysis of green bug resistance in barley. Crop Sci 13:684–685

Gaskill JO, Mumford DL, Ruppel EG (1970) Preliminary report on breeding for combined resistance to leaf spot curly top, and *Rhizoctonia*. J Am Soc Sugar Beet Technol 16:207–213

Gäumann E (1964) Weitere Untersuchungen über die chemische Infektabwehr der Orchideen. Phytopathol Z 49:211–232

Geigert J, Stermits FR, Johnson D, Maag DD, Johnson DK (1973) Two phytoalexins from sugarbeet *(Beta vulgaris)* leaves. Tetrahedron 29:2703–2706

Gentry HS (1969) Origin of the common bean, *Phaseolus vulgaris*. Econ Bot 23:55–69

Genung WG, Green VE (1962) Insects attacking soybeans with emphasis on varietal susceptibility. Soils Crop Sci Soc Florida 22:138–142

Gerechter-Amitai ZK, Wahl I, Vardi A, Zohary D (1971) Transfer of stem rust seedling resistance from wild diploid Einkorn to tetraploid durum wheat by means of a triploid hybrid bridge. Euphytica 20:281–285

Gibson RW (1976a) Glandular hairs are a possible means of limiting aphid damage to the potato crop. Ann Appl Biol 82:143–146

Gibson RW (1976b) Glandular hairs on *Solanum polyadenium* lessen damage by the Colorado beetle. Ann Appl Biol 82:147–150

Gibson RW (1979) The geographical distribution, inheritance and pest resting properties of sticky tipped foliar hairs on potato species. Potato Res 22:223–236

Gill KS, Nanda GS, Singh G (1975) Studies of multilines in wheat (*T. aestivum* L.). Stability of multilines of Kalyansona and PV 18 cultivars. Crop Improv 2(1, 2):9–16

Gill KS, Kumar V, Nanda GS (1977) Multilines in wheat. 4, Performance of the multilines and components of PV 18. Ind J Genet Plant Breed 37:388–394

Gill KS, Aujla SS, Sharma YR, Nanda GS, Singh G (1978) Studies on multilines in wheat *(T. aestivum)*. Seedling and adult plant reaction of components of Kalyansona and PV 18 multilines against yellow rust. Cereal Res Commun 6:167–173

Gill KS, Nanda GS, Singh G, Aujla SS (1979) Multilines in wheat – A review. Ind J Genet Plant Breed 39:30–37

Gill KS, Nanda GS, Singh G, Aujla SS (1980) Studies on multilines in wheat (*Triticum aestivum* L.). 12. Breeding of a multiline variety by convergence of breeding lines. Euphytica 29:125–128

Gill SS, Pandey S, Tripathi BK (1984) Sources of resistance to red rot of sugarcane caused by *Physalospora tucumanensis*. Ind J Pathol 2:43–45

Gilman DF, McPherson RM, Newsom LD, Herzog DC, Williams C (1982) Resistance in soybeans to the southern green stinck bug. Crop Sci 22:573–576

Glazener JA, Wouters CH (1981) Detection of rishitin in tomato fruits after infection with *Botrytis cinerea*. Physiol Plant Pathol 19:243–248

Glover C, Melton B (1966) Inheritance patterns of spotted alfalfa aphid resistance in *Zea* plants. N.M. Agric Exp Stn Res Rep 127:4

Glover DV, Stanford EH (1966) Tetrasomic inheritance of resistance in alfalfa to the pea aphid. Crop Sci 6:161–165

Gokhale VP, Patel MK (1952) Effect of stem rust, particularly on grain weight of some improved Indian Wheats. Ind Phytopathol 5:89–103

Goldschmidt V (1928) Vererbungsversuche mit den biologischen Arten des Antherenbrandes *(U. violacea)*. Ein Beitrag zur Frage der parasitären Specialisierung. Z Bot 21:1

Gough FJ, Merkle OG, Statler GD (1974) Inheritance of stem and leaf rust resistance in skorospelka 3b wheat. Crop Sci 14:330–332

Goulden CH, Neatby KW, Welsh JN (1928) The inheritance of resistance to *Puccinia graminis tritici* in cross between two varieties of *Triticum vulgare*. Phytopathology 18:631–658

Gracen VE, Forster JJ, Sayre KD, Grogan CO (1971) Rapid method for selecting resistant plants for control of southern corn leaf blight. Plant Dis Rep 55:469–470

Grafius JE (1966) Rate of change of lodging resistance, yield, and test weight in varietal mixtures of oats, *Avena sativa* L. Crop Sci 6:369–370

Graham JH, Hill RR Jr, Barnes DK, Hanson CH (1965) Effect of three cycles of selection for resistance to common leaf spot in alfalfa. Crop Sci 5:171–173

Graham KM (1955) Distribution of physiological races of *Phytophthora infestans* (Mont.) de Bary in Canada. Am Potato J 32:277–288

Grama A, Gerechter-Amitai ZK (1974) Inheritance of resistance to stripe rust *(Puccinia striiformis)* in crosses, between wild emmer *(Triticum dicocoides)* and hexaploid wheat. II. *Triticum aestivum*. Euphytica 23:393–398

Green GJ (1964) A color mutation, its inheritance, and the inheritance of pathogenicity in *Puccinia graminis*. Pers. Can J Bot 42:1653–1664

Green GJ (1966) Selfing studies with races 10 and 11 of wheat stem rust. Can J Bot 44:1255–1260

Green GJ (1971) Hybridization between *Puccinia graminis tritici* and *Puccinia graminis secalis* and its evolutionary implications. Can J Bot 49:2089–2095

Green GJ, Kirmani MAS (1969) Somatic segregation in *Puccinia graminis* f.sp. *avenae*. Phytopathology 59:1106–1108

Green GJ, McKenzie RIH (1967) Mendelian and extrachromosomal inheritance of virulence in *Puccinia graminis* f.sp. *avenae*. Can J Genet Cytol 9:785–793

Griffiths DJ, Holden JHW, Jones JM (1957) Investigations on resistance of oats to stem eelworm, *Ditylenchus dipsaci* Kuhn. Ann Appl Biol 45:709–720

Grisebach H, Ebel J (1978) Phytoalexins, chemical defense substances of higher plants. Angew Chem Int Ed Engl 17:635–647

Grogan CO (1971) Multiplasm a proposed method for the utilization of cytoplasms in pest control. Plant Dis Rep 55:400–401

Gross G (1977) Phytoalexine und verwandte Pflanzenstoffe. Prog Chem Org Nat Prod 34:187–247

Gunathilagaraj K, Chelliah S (1985) Components of resistance to the whitebacked planthopper, *Sogatella furcifera* (Horvath), in some rice varieties. Trop Pest Manage 31:38–46

Gunawardena SDIE, Navaratne SK, Ganashan P (1971) Rice breeding with induced mutations in Ceylon. In: Proc Symp Rice Breed Induced Mutat, vol III. Vienna Int At Energy Ag Tech Rep Ser No 131:29–33

Gundy SD van (1972) Non chemical control of nematodes and root-infecting fungi. In: Pest control strategies for the future. Natl Acad Sci, Washington DC, pp 312–329

Gupta AK, Saini RG, Gupta S, Malhotra S (1984) Genetic analysis of two wheat cultivars, 'Sonalika' and 'WL 711' for reaction to leaf rust *(Puccinia recondita)*. Theor Appl Genet 67:215–217

Gupta RP, Singh A (1982) Slow rusting in wheat infected with brown rust. Ind J Genet Plant Breed 42:172–182

Guthrie WD (1969) European corn borer infestigations. Eur Corn Borer Lab, Ankeny, Iowa

Guthrie WD, Walter EV (1961) Corn ear worm and European corn borer resistance in sweet corn inbred lines. J Econ Entomol 54:1248–1250

Haberlach GT, Budde AD, Sequeira L, Helgeson JP (1978) Modification of disease resistance of tobacco callus tissue by cytokinins. Plant Physiol 62:522–525

Habgood RM (1970) Designation of physiological races of plant pathogens. Nature (London) 227: 1268–1269

Hackerott JL, Sorensen EL, Harvery TL, Ortman EE, Painter RH (1963) Reactions of alfalfa varieties to pea aphids in the field and green house. Crop Sci 3:298–301

Haddad NI, Muehlbauer FJ, Hampton RO (1978) Inheritance of resistance to pea seed-borne mosaic virus in lentils. Crop Sci 18:613–615

Hadwiger LA, Beckmann JM (1980) Chitosan as a component of Pea-*Fusarium solani* interactions. Plant Physiol 66:205–211

Hagborg WAF (1979) A method of estimating the resistance of wheat varieties to a bacterial disease. In: Proc Symp 5th Int Genet Wheat New Delhi 2:1102–1104

Hagedorn DJ, Gritton ET (1973) Inheritance of resistance to the pea seed-borne mosaic virus. Phytopathology 63:1130–1133

Hagen KS (1958) Honeydew as an adult fruitfly diet affecting reproduction. In: Proc Int Congr Entomol 10:25–30

Hagges ME, Dyck PL (1973) The inheritance of leaf rust resistance in four common wheat varieties possessing genes at or near the Lr locus. Can Genet Cytol 15:127–134

Hahn SK (1968) Resistance of barley (*Hordeum vulgare* L. emend. Lam.) to cereal leaf beetle (*Oulema melanopus* L.). Crop Sci 8:461–464

Halalli MS, Gowda BTS, Kulkarni KA, Goud JV (1982) Inheritance of resistance to shootfly (*Atherigona soccata* Rond.) in *sorghum* (*Sorghum bicolor* (L.) Moench). SABRAO J 14:165–170

Hallman GJ, Edwards GR, Foster JE (1977) Soybean cultivars evaluated for resistance to Mexican bean beetle in southern Indiana. J Econ Entomol 70:316–318

Hanna WW, Wells HD, Burton GB (1985) Dominant gene for rust resistance in pearl millet. J Hered 76:134

Hanover JW (1975) Physiology of tree resistance to insects. Annu Rev Entomol 20:75–79

Hanson CH, Norwood BL, Blickenstaff CC, Van Denburgh RS (1963) Recurrent phenotypic selection for resistance to potato leaf hopper yellowing in alfalfa. Agron Abstr, p 80

Hanson CH, Busbice JH, Hill RR Jr, Hunt OJ, Oakes AJ (1972) Directed mass selection for developing multiple pest resistance and conserving germplasm in alfalfa. J Envrion Qual 1:106–111

Hardegger E, Biland HR, Corrodi H (1963) Synthese von 2,4-Dimethoxy-6-hydroxy-phenanthren und Konstitution des Orchinols Helv. Chim Acta 46:1354–1360

Harder DE, McKenzie RIH, Martens JW (1984) Inheritance of adult plant resistance to crown rust in an accession of *Avena sterilis*. Phytopathology 74:352–353

Hare WW (1965) The inheritance of resistance of plants to nematode. Phytopathology 55:1162–1167

Hargreaves JA, Mansfield JW, Rossall S (1977) Changes in phytoalexin concentrations in tissues of the broad bean plant (*Vicia faba* L.) following inoculation with species of *Botrytis*. Physiol Plant Pathol 11:227–242

Harlan JR (1975a) Geographical patterns of variation in some cultivated species. J Hered 66:182–191

Harlan JR (1975b) Our vanishing genetic resources. Science 188:618–621

Harlan JR (1976) Genetic resources in wild relatives of crops. Crop Sci 16:329–333

Harlan JR, Zohary D (1966) Distribution of wild wheats and barley. Science 153:1074–1080

Harrington CD (1943) The occurrence of physiological races of pea aphid. J Econ Entomol 36: 118–119

Harris-Warrich RM, Lederberg J (1978) Interspecific transformation in *Bacillus* mechanism of heterologous intergenote transformation. J Bacteriol 133:1246–1253

Hartley MJ, Williams PG (1971) Genotypic variation within a phenotype as a possible basis for somatic hybridization in rust fungi. Can J Bot 49:1085–1087

Hartwig EE, Bromfield KR (1983) Relationships among three genes conferring specific resistance to rust in soybeans. Crop Sci 23:237–239

Harvey TL, Hackerott HL, Sorensen E (1972) Pea aphid-resistant alfalfa selected in the field. J Econ Entomol 65:1661–1663

Harwood RR, Grandados RY, Jamornman S, Granados RG (1973) Breeding for resistance to sorghum shootfly in Thailand. In: Control of sorghum shootfly. IBH, Oxford New Delhi

Hatchett JH, Gallun RL (1970) Genetics of the ability of the hessian fly, *Mayetiola destructor*, to survive on wheats having different genes for resistance. Ann Entomol Soc Am 63:1400–1407

Hatchett JH (1969), Race E sixth race of the hessian-fly, *Mayetiola destructor*, discovered in Georgia wheat fields. Ann Entomol Soc Am 62:677–679

Hawkes JG (1958) Significance of wild species and primitive forms for potato breeding. Euphytica 7:257–270

Hayden EB (1956) Differences in infectibility among spring wheat varieties exposed to spore showers of race 15B of *Puccinia graminis* var. *tritici*. Phytopathology 46:14 (Abstr)

Hayes HK, Immer FR (1942) Methods of plant breeding. McGraw Hill, New York

Hayes JD (1973) Prospects for controlling cereal disease by breeding for increased levels of resistance. Ann Appl Biol 75:140–144

Heath MC, Heath IB (1971) Ultrastructure of an immune and a susceptible reaction of cowpea leaves to rust infection. Physiol Plant Pathol 1:277–278

Heijbroek W, Roelands AJ, De Jong JH (1983) Transfer of resistance to beet cyst nematode from *Beta patellaris* to sugarbeet. Euphytica 32:287–298

Helgeson JP (1983) Studies of host-pathogen interactions in vitro. In: Use of tissue culture and protoplasts in plant pathology. Academic Press, Sydney, pp 9–38

Herrera-Estrella L, De Block M, Messens E, Hernalsteens JP, Von Montagen M, Schell J (1983) Chimeric genes as dominant selectable markers in plant cells. EMBO J 2:987–995

Higgins VT, Smith DG (1972) Separation and identification of two pterocarpanoid phytoalexins produced by red clover leaves. Phytopathology 62:235–238

Hijano EH, Barnes DK, Frosheiser FI (1983) Inheritance of resistance to *Fusarium* wilt in alfalfa. Crop Sci 23:31–34

Hill HJ, West SH (1982) Fungal penetration of soybean seed through pores. Crop Sci 22:602–605

Hill RR Jr, Hanson CH, Busbice TH (1969) Effect of four recurrent selection programs on two alfalfa populations. Crop Sci 9:363–365

Ho KM, Jones GE (1980) Mingo barley. Can J Plant Sci 60:279–280

Hodgson WH (1961) Laboratory testing on the potato for partial resistance to *Phytophthora infestans*. Am Potato J 38:259–264

Hogen-Esch JA, Zingstra H (1957) Geniteurslijst voor aartappelrassen 1957, 147 pp. Comissieter bevordering van het kweken en het onderzoek van nieuwe aartapplrassen. Wegeningen, Netherlands

Hollander J den, Pathak PK (1981) The genetics of the biotypes of the rice brown plant hopper, *Nilaparvata lugens*. Entomol Exp Appl 29:76–86

Holmes FO (1938) Inheritance of resistance to tobacco mosaic disease in tobacco. Phytopathology 28:553–561

Holmes FO (1939) The Chilean tomato, *Lycopersicon chilense* as a possible source of disease resistance. Phytopathology 29:215–216

Holmes ND (1984) The effect of light on the resistance of hard red spring wheats to the wheat stem sawfly, *Cephus cinctus* (Hymenoptera:Cephidae). Can Entomol 116:677–684

Holmes ND, Peterson LK (1957) Effect of continuous rearing on Rescue wheat on survival of the wheat stem sawfly, *Cephus cinctus* Nort. (Hymenoptera:Cephidae). Can Entomol 89:363–365

Holmes ND, Peterson LK (1962) Resistance of spring wheats to the wheat stem sawfly, *Cephus cinctus* Nort. (Hymenoptera:Cephidae). II. Resistance to the larvae. Can Entomol 94:348–364

Holton CS, Halisky CS (1960) Dominance of avirulence and monogenic control of virulence in race hybrids of *Ustilago avenae*. Phytopathology 50:766–770

Holton CS, Hoffman JA, Duran R (1968) Variation in the smut fungi. Annu Rev Phytopathol
    6:213–242
Hooker AL (1963) Monogenic resistance in *Zea mays* L. to *Helminthosporium turcicum*. Crop
    Sci 3:381–383
Hooker AL (1967) The genetics and expression of resistance in plants to rusts of the genus *Puc-
    cinia*. Annu Rev Phytopathol 5:163–182
Hooker AL (1967) Inheritance of mature plant resistance to rust in corn. Phytopathology 57:
    815 (Abstr)
Hooker AL (1973) Maize. In: Nelson RR (ed) Breeding plants for disease resistance. State Univ
    Press, London, pp 132–152
Hooker AL (1978) Genetics of disease resistance in maize. In: Walden DB (ed) Maize breeding and
    genetics. Wiley, New York, pp 319–322
Hooker AL, Kim SK (1973) Monogenic and multigenic resistance to *Helminthosporium turcicum*
    in corn. Plant Dis Rep 57:586–589
Hooker AL, Russell WA (1962) Inheritance of resistance to *Puccinia sorghi* in six corn inbred lines.
    Phytopathology 52:122–128
Hooker AL, Saxena KMS (1967) Apparent reversal of dominance of a gene in corn for resistance to
    Puccinia sorghi. Phytopathology 57:1372–1374
Hooker AL, Tsung YK (1980) Relationship of dominant genes in corn for chlorotic lesion resistance
    to *Helminthosporium turcicum*. Plant Dis Rep 4:387–388
Hooker AL, Smith DR, Lim SM, Beckett JB (1970) Reaction of corn seedlings with male-sterile
    cytoplasm to *Helminthosporium maydis*. Plant Dis Rep 54:708–712
Horber E (1979) Types and classification of resistance. In: Maxwell FG, Jennings PR (eds) Breed-
    ing plants resistant to insects. Wiley Interscience, pp 15–21
Horino O, Mew TW, Yamada T (1981) The effect of temperature on the development of bacterial
    leaf blight on rice. Ann Phytopathol Soc Jpn 48:72–75
Howard HW (1968) The relation between resistance genes in potatoes and pathotypes of potato-
    root eelworm *(Heterodera rostochiensis)*, wart disease *(Synchytrium endobioticum)* and potato
    virus. Xth Int Congr Plant Pathol (Abstr Ist) 92, London
Howard HW, Fuller JM (1975) Testing potatoes bred from *Solanum vernei* for resistance to the
    white potato cyst-nematode, *Heterodera pallida*. Ann Appl Biol 81:75–78
Hughes GR, Hooker AL (1971) Genes conditioning resistance to northern leaf blight in maize. Crop
    Sci 11:180–184
Hunt JS, Barnes MF (1982) Molecular diversity and plant disease resistance: An electrophoretic
    comparison of near isogenic lines of wilt resistant or susceptible *Pisum sativum* L. Cv. William
    Massey. Euphytica 31:341–348
Hutton EM, Peak AR (1949) Spotted wilt resistance in tomato. J Aust Inst Agric Sci 15:32
Hwang BK, Heitefuss R (1982) Characterization of adult plant resistance of spring barley to powdery
    mildew *(Erysiphe graminis* f.sp. *hordei)*. I. Race specificity and expression of resistance. Phyto-
    pathol Z 104:168–178
Hyde PM (1982) Temperature sensitive resistance of the wheat cultivar Maris Fundin to *Puccinia
    recondita*. Plant Pathol 31:25–30
Ibrahim G, Owen H, Ingham JL (1982) Accumulation of medicarpin in broad bean roots: A pos-
    sible factor in resistance to *Fusarium oxysporum* root rot. Phytopathol Z 105:20–26
Ikeda R, Kaneda C (1982) Genetic studies on brown planthopper resistance of rice in Japan. JARQ
    16:1–5
Indira S, Rana BS, Rao NGP (1983) Genetics of host plant resistance to charcoal rot in sorghum.
    Ind J Genet Plant Breed 43:472–477
Ingham JL, Miller RL (1973) Sativin: an induced isoflavin from the leaves of *Medicago sativa* L.
    Nature (London) 242:125–126
Innes NL (1964) Resistance conferred by new gene combinations to bacterial blight of cotton.
    Euphytica 13:33–43
Innes NL, Brown SJ, Walker JT (1974) Genetical and environmental variation for resistance to
    bacterial blight of upland cotton. Heredity 32:53–72
IRAT (1973) Inst Rech Agron Trop Cult Viv, Bouake, Ivory Coast. Annu Rep Rice Plant Pathol
    1972

IRAT (1976) Inst Rech Agron Trop Cult Viv, Bouake, Ivory Coast. Annu Rep Rice Plant Pathol 1974, 1975

IRAT (1977) Inst Rech Agron Trop Cult Viv, Bouake, Ivory Coast. Annu Rep Rice Plant Pathol 1976

IRRI (1966) Int Rice Res Inst Philippines. Annu Rep Plant Pathol, p 94

IRRI (1970) Int Rice Res Inst, Philippines. Annu Rep 1969

IRRI (1974) Int Rice Res Inst, Philippines. Annu Rep

IRRI (1978) Int Rice Res Inst, Philippines. Res Highlights 1977, p 28

IRRI (1980) Int Rice Res Inst, Philippines. Res Highlights 1979, pp 28–29

Irwin JAG, Maxwell DP, Bingham ET (1981) Inheritance of resistance to *Phytophthora megasperma* in diploid alfalfa. Crop Sci 21:271–276

Jackai LE (1982) A field screening technique for resistance of cowpea *(Vigna unguiculata)* to the pod-borer, *Maruca testulalis* (Geyer) (Lepidoptera:Pyralidae). Bull Entomol Res 72:145–156

Jain KBL, Upadhyay MK (1974) Genetics of loose smut resistance in barley. Ind J Genet Plant Breed 34:161–1163

Jain R, Gandhi SM (1978) Genetics of stem rust resistance in wheat variety Timgalen. Ind J Genet Plant Breed 38:252–257

Jain R, Gandhi SM (1980) Inheritance of stem rust resistance in wheat varieties TOBARI and ZAMBESI. Ind J Genet Plant Breed 40:602–607

Jain R, Gandhi SM (1983) Genes for stem rust resistance in wheat variety Raj 848. Ind J Genet Plant Breed 43:356–360

Jambhale ND, Nerkar YS (1981) Inheritance of resistance to Okra yellow vein moaic disease in interspecific crosses of *Abelmoschus*. Theor Appl Genet 60:313–316

James RV, Williams PH (1980) Clubroot resistance and linkage in *Brassica campestris*. Phytopathology 70:776–779

James WC (1974) Assessment of plant disease and losses. Annu Rev Phytopathol 12:27–48

Jan J de, Honma S (1976) Inheritance of resistance to *Corynebacterium michiganense* in the tomato. J Hered 67:79–84

Jarvis JS (1970) Relative injury to some Cruciferous oilseeds by the turnip aphid. J Econ Entomol 63:1498–1502

Jayraj D, Murty VVS (1983) Inheritance of resistance to white backed planthopper, *Sogatella furcifera* (Horvath) in rice. In: 15th Int Congr Genet Abstr No 1309:724

Jeger MJ, Jones DG, Griffiths E (1983) Components of partial resistance of wheat seedlings to *Septoria nodorum*. Euphytica 32:575–584

Jenkins JN (1981) Breeding for insect resistance. In: Frey KJ (ed) Plant breeding, vol II. Iowa State Univ Press, Iowa Ames, USA, pp 291–308

Jenkins JN, Maxwell FG, Parrott WL (1964) A technique for measuring certain aspects of antibiosis in cotton to the boll weevil. J Econ Entomol 57:679–681

Jenkins JN, Maxwell FG, Parrott WL, Buford WT (1969) Resistance to boll weevil (*Anthonomus grandis* Boh.) oviposition in cotton. Crop Sci 9:369–372

Jenkins MT (1940) The segregation of genes affecting yield of grain in maize. J Am Soc Agron 32: 55–63

Jenkins MT, Roberts AL, Findley WR Jr (1954) Recurrent selection as a method for concentrating genes for resistance to *Helminthosporium turcicum* leaf blight in corn. Agron J 46:89–94

Jenng HR (1975) Inheritance of resistance of a rice mutant. MR 515–17 to *Pyricularia oryzae*, race IA-65. SABRAO J 7:61–64

Jennings CW, Russell WA, Guthrie WD (1974) Genetics of resistance in maize to first and second-brood European corn borer. Crop Sci 14:394–398

Jennings PR, Pineda AT (1970) Screening rice for resistance to the planthopper *Sogatodes orizicola* (Muir). Crop Sci 10:687–689

Jensen NF (1952) Intra-varietal diversification in oat breeding. Agron J 44:30–34

Jensen NF (1965) Population variability in small grains. Agron J 57:153–162

Jensen NF (1966) Broadbase hybrid wheats. Crop Sci 6:376–377

Jensen NF, Kent GC (1963) New approach to an old problem in oat production. Farm Res 29: 4–5

Jinks JL (1966) Extranuclear inheritance. In: Ainsworth GC, Sussman AS (eds) The fungi: an advanced treatise, vol II. Academic Press, London New York, pp 619–660

John WVD, Sam GT, Maxwell JD (1971) Resistance in soybeans to the Mexican bean beetle. I. Sources of resistance. Crop Sci 11:572–573

Johnson AG, Crute IR, Gordon PL (1977) The genetics of race specific resistance in lettuce *(Lactuca sativa)* to downy mildew *(Bremia lactucae)*. Ann Appl Biol 86:87–103

Johnson C, Brannon DR, Kuc J (1973) Xanthotoxin. Phytoalexin of *Pastinaca sativa* root. Phytochemistry 12:2961–2962

Johnson DA, Wilcoxson RD (1978) Components of slow rusting in barley infected with *Puccinia hordei*. Phytopathology 68:1470–1474

Johnson DA, Wilcoxson RD (1979) Inheritance of slow rusting of barley infected with *Puccinia hordei* and selection of latent period and number of uredia. Phytopathology 69:145–151

Johnson DR, Campbell WV, Wynne JC (1982) Resistance of peanuts to the two-spotted spider mite (Acari:Tetranychidao). J Econ Entomol 75:1045–1047

Johnson R (1966) The substitution of a chromosome from *Agropyron elongatum* for chromosomes of hexaploid wheat. Can J Genet Cytol 8:279–292

Johnson R (1978) Practical breeding for durable resistance to rust disease in self-pollinating cereals. Euphytica 27:529–540

Johnson T, Green GJ, Samborski DJ (1967) The world situation of the cereal rusts. Annu Rev Phytopathol 5:183–200

Johnston TD (1974) Transfer of disease resistance from *Brassica campestris* L. to rape (*B. napus* L.). Euphytica 23:681–683

Jones DF, Parrott DM (1965) The genetic relationship of pathotypes of *Heterodera rostochiensis* Woll. Which reproduce on hybrid potatoes with genes for resistance. Ann Appl Biol 56:27–36

Jones FGW (1974) Host-parasite relationships of potato cystnematodes: a speculation arising from gene-for-gene hypothesis. Nematologia 20:437–443

Jones IT (1978) Components of adult plant resistance to powdery mildew (*Erysiphe graminis* f.sp. *avenae*) in oats. Ann Appl Biol 90:233–239

Jones IT, Pickering RA (1978) The mildew resistance of *Hordeum bulbosum* and its transferance into *H. vulgare* genotypes. Ann Appl Biol 88:295–298

Jones JE, Newson LD, Tipton KW (1964) Differences in boll weevil infestation among several biotypes of upland cotton. Proc Ann Cotton Improv Conf 16:48–55

Jones JP, Hartwig E (1959) A simplified method for field inoculation of soybeans with bacteria. Plant Dis Rep 43:946

Jordan VWL (1973) A procedure for the rapid screening of strawberry seedlings for resistance to *Verticillium* wilt. Euphytica 22:367–372

Jørgensen JH (1971) Comparison of induced mutant genes with spontaneous genes in barley conditioning resistance to powdery mildew. In: Proc Symp Mutat Breed Dis Resist. Int At Energy Ag Vienna, pp 117–124

Joshi LM, Adlakha KL (1974) Sources of resistance to *Helminthosporium sativum*. Ind J Genet Plant Breed 34:368–370

Joshi LM, Gera SD, Saari EE (1973) Extensive cultivation of Kalyansona and disease development. Ind Phytopathol 26:371–373

Joshi LM, Singh DV, Srivastava KD, Wilcoxson RD (1983) Karnal Bunt: A minor disease that is now a threat to wheat. Bot Rev 49:309–330

Kalode MB, Pant NC (1967) Studies on the aminoacids, nitrogen, sugar and moisture content of maize and sorghum varieties and their relation to *Chilo zonellus* (Swin.) resistance. Ind J Entomol 29:139–144

Kaneko K, Aday BA (1980) Inheritance of resistance to Philippine downy mildew of maize, *Peronosclerospora philippinensis*. Crop Sci 20:590–594

Kannaiyan J, Nene YL, Raju TN, Sheila VK (1981) Screening for resistance to *Phytophthora* blight of pigeon-pea. Plant Dis 65:61–62

Kao KN, Knott DR (1969) The inheritance of pathogenicity in races 111 and 29 of wheat stem rust. Can J Genet Cytol 11:266–274

Katsui N, Murai A, Takasugi M, Imaizumi K, Masumane T, Toriayama K (1968) The structure of rishitin a new antifungal compound from diseased potato tubers. Chem Commun 43–44

Kauffman HE, Reddy APK, Hsieh SPY, Merca SD (1973) An improved technique for evaluating resistance of rice varieties to *Xanthomonas oryzae*. Plant Dis Rep 57:537–541

Kaur S, Padmanabhan SY, Kaur P (1977) Induction of resistance to blast disease *(Pyricularia oryzae)* in the high yielding variety, Ratna (IR 8 × TKM 6). In: Proc Symp Induced Mutat Against Plant Dis. Int At Energy Ag, Vienna, pp 147–156

Kaushal K, Upadhyaya YM (1983) Field studies on genetic diversity for field resistance to black and brown rust in crosses between resistant parents. Ind J Genet Plant Breed 43:414–417

Keaster AJ, Zuber MS, Straub RW (1972) Status of resistance to corn earworm. Proc North Cent Branch Entomol Soc Am 27:95–98

Keen NT (1972) Accumulation of wyerone in broad been and demethylhomopterocarpin in jackbean after inoculation with *Phytophthora megasperma* var. *sojae*. Phytopathology 62:1365–1367

Keen NT (1975) Isolation of phytoalexins from germinating seeds of *Cicer arietinum*, *Vigna sinensis*, *Arachis hypogaea* and other plants. Phytopathology 65:91–92

Keen NT, Horsch R (1972) Hydroxy-phaseollin production by various soybean tissues: A warning against the use of unnatural host-parasite systems. Phytopathology 62:439–442

Keep E (1972) Variability in wild raspberry. New Phytol 71:915

Keep E, Knight RL (1967) A new gene from *Rubus occidentalis* L. for resistance to strain 1, 2, and 3 of the *Rubus* aphid, *Amphorophora rubi* kalt. Euphytica 16:209–214

Kellner E (1892) Report on grains. Calif Agric Exp Stn Rep 1891–92:138–144

Kennedy GG (1976) Host plant resistance and the spread of plant viruses. Environ Entomol 5:827–832

Kennedy GG, Schaefers GA (1974) Evaluation of immunity in red raspberry to *Amphorophora agathonica* via cuttings. J Econ Entomol 67:311–312

Kennedy GG, Schaefers GA (1975) Role of nutrition in the immunity of red raspberry to *Amphorophora agathonica* Hottes. Environ Entomol 4:115–119

Kerber EF, Dyck PL (1969) Inheritance in hexaploid wheat of leaf rust resistance and other characters derived from *Aegilops squamosa*. Can J Genet Cytol 11:639–647

Kerber EF, Dyck PL (1973) Inheritance of stem rust resistance transferred from diploid wheat *(Triticum monococcum)* to tetraploid and hexaploid wheat and chromosome location of the gene involved. Can J Genet Cytol 19:639–647

Kerr EA, Bailey DL (1964) Resistance to *Cladosporium fulvum* Ckc obtained from wild species of tomato. Can J Bot 42:1541–1554

Kesavan V, Choodhury B (1977) Screening for resistance to *Fusarium* wilt of tomato. SABRAO J 9:57–65

Khalf-Allah AM, Paris FS, Nassar SH (1973) Inheritance and nature of resistance to cucumber mosaic virus in cowpea, *Vigna sinensis*. Egypt J Genet Cytol 2:274–282

Khambanonda P (1971) Rice breeding with induced mutations in Thailand: review of studies made over a five year period. In: Proc Symp Rice Breed Induced Mutat, vol III. Vienna Int At Energy Ag, Tech Rep Ser No 131:61–68

Khan AQ, Paliwal RL (1979) Inheritance of stalk rot resistance in maize. Ind J Genet Plant Breed 39:139–145

Khandelwal RC, Nath P (1978) Inheritance of resistance to fruit fly in watermelon. Can J Genet Cytol 20:31–34

Khandelwal RC, Nath P (1979) Evaluation of cultivars of watermelon for resistance to fruit fly. Egypt J Genet Cytol 8:46–56

Khush GS (1971) Rice breeding for disease and insect resistance at IRRI. Oryza (Suppl) 8(2):111–119

Khush GS (1977a) Disease and insect resistance in rice. Adv Agron 29:265–333

Khush GS (1977b) Breeding for resistance in rice. Ann NY Acad Sci 287:296–308

Khush GS (1979) Genetics of and breeding for resistance to the brown plant hopper. In: Proc Symp Brown Plant Hopper. Threat to rice production in Asia. IRRI, Philippines, pp 321–332

Khush GS, Beachell HM (1972) Breeding for disease and insect resistance at IRRI. In: Proc Symp Rice Breed. IRRI, Philippines, p 738

Khush GS, Ling KC (1974) Inheritance of resistance to grassy stunt virus and its vector in rice. J Hered 65:134–136

Kilen TC, Keeling BL (1981) Genetics of resistance to *Phytophthora* rot in soybean cultivar PI 171442. Crop Sci 21:873–875

Kilen TC, Keeling BL, Hartwig EE (1985) Inheritance of reaction to stem canker in soybean. Crop Sci 25:50–51

Kim BS, Hartmann RW (1985) Inheritance of a gene (Bs3) conferring hypersensitive resistance to *Xanthomonas campestris* pv. *vesicatoria* in pepper *(Capsicum annum)*. Plant Dis 69:233–235

Kim SK, Brewbaker JL (1977) Inheritance of general resistance in maize to *Puccinia sorghi*. Crop Sci 17:456–461

Kimber G (1967) The incorporation of the resistance of *Aegilops ventricosa* to *Cercosporella herpotrichoides* into *Triticum aestivum*. J Agric Sci 68:373–376

Kindler SD, Schalk JM (1975) Frequency of alfalfa plants with combined resistance to the pea aphid and spotted alfalfa aphid in aphid resistant cultivars. J Econ Entomol 68:716–718

King PJ, Potrykus I, Thomas E (1978) In-vitro genetics of cereals: Problems and perspectives. Physiol Veg 16:381–399

Kiraly A, Farkas GL (1962) Relation between phenol metabolism and stem rust resistance in wheat. Phytopathology 52:657–664

Kiraly Z (1971) Artificial inoculation methods for screening resistant types. In: Proc Symp Mutat Breed Dis Resist. Int At Energy Ag, Vienna, pp 201–205

Kiraly Z, Klement Z, Solymosy F, Voros J (1970) Methods in plant pathology (with special reference to breeding for disease resistance). Akad Kiado, Budapest

Kiraly Z, Barna B, Ersek T (1972) Hypersensitivity as a consequence, not the cause, of plant resistance to infection. Nature (London) 239:456–458

Kirste (1958) Ergebnisse von Krautfäule-Spritzversuchen. Kartoffelbau 9:114–115

Kishaba AN, Manglitz GR (1965) Non preference as a mechanism of sweet clover and alfalfa resistance to the sweet clover aphid and the spotted alfalfa aphid. J Econ Entomol 58:566–569

Kishaba AN, Bohn GW, Toba EH (1971) Resistance to *Aphis gossypii* in muskmelon. J Econ Entomol 64:935–937

Kiyosawa S (1984) Population genetics of rice blast relationships. In: Chopra VL, Joshi BC, Sharma RP, Bansal HC (eds) Genetics, new frontiers. Proc XVth Int Congr Genet. IBH, Oxford, pp 119–130

Kiyosawa S, Kushibuchi K, Watanabe S (1975) Recent trends and problems in breeding for blast resistant varieties and its investigation. Nogyo Oyobi Engei (Agric Hortic) 50:25–30

Klashorst GV, Tingey WM (1979) Effect of seedling age, environmental temperature and foliar total glycoalkoids resistance of five *Solanum* genotypes to the potato leaf hopper. Environ Entomol 8:690–693

Klement Z (1968) Phytogenicity factors in regards to relationships of phytopathogenic bacteria. Phytopathology 58:1218–1221

Klement Z, Goodman RN (1967) The hypersensitive reaction to infection by bacterial plant pathogens. Annu Rev Phytopathol 5:17–44

Klun JA, Brindley TA (1966) Role of 6-methoxybenzoxalinone in inbred resistance of host plant (Maize) to the first brood larvae of European corn borer. J Econ Entomol 59:711–718

Klun JA, Tipton CL, Brindley TA (1967) 2,4-dihydroxy-7-methoxy-1, 4-benzoxazin-3-one (DIMBOA), an active agent in the resistance of maize to the European corn borer. J Econ Entomol 60:1529–1533

Klun JA, Guthrie WD, Hallauer AR, Russell WA (1970) Genetic nature of the concentration of 2,4-dihydroxy-7-methoxy-2H-1, 4 benzoxazin 3 (4H)-one and resistance to the European corn borer in a diallel set of eleven maize inbreds. Crop Sci 10:87–90

Knapp JL (1966) A comparison of resistant and susceptible dent corn single crosses to damage by the corn ear worm, *Heliothis zea* Boddie. Diss Abstr 26:4918

Knapp JL, Maxwell FG, Douglas WA (1967) Possible mechanisms of resistance of dent corn to the corn ear worm. J Econ Entomol 60:33–36

Knight RL (1954) Cotton breeding in Sudan. Part I and II. Egyptian cotton. Emp J Exp Agric 22:68–92

Knight RL, Briggs JB, Keep E (1960) Genetics of resistance to *Amphorophora rubi* (Kalt.) in the raspberry. Genet Res 1:319–331

Knight TA (1799) Philos Trans R Soc London Ser B, p 192

Knott DR (1961) The inheritance of rust resistance. VI. The transfer of stem rust resistance from *Agropyron elongatum* to common wheat. Can J Plant Sci 41:109–123

Knott DR (1964) The effect on wheat of an *Agropyron* chromosome carrying rust resistance. Can J Genet Cytol 6:500–507

Knott DR (1968) The inheritance of resistance to stem rust races 56 and 15B-11 (Can.) in the wheat varieties Hope and H44. Can J Genet Cytol 10:311–320

Knott DR (1971a) The transfer of genes for disease resistance from alien species to wheat by induced translocations. In: Proc Symp Mutat Breed Dis Resist. Int At Energy Ag, Vienna, pp 67–77

Knott DR (1971b) Genes for stem rust resistance in wheat varieties Hope and H44. Can J Genet Cytol 13:186–188

Knott DR (1979) The inheritance of general resistance to stem rust in wheat. In: Proc Symp 5th Int Wheat Genet. New Delhi 2:1079–1086

Knott DR (1981) The effects of genotype and temperature on the resistance to *Puccinia graminis tritici* controlled by the gene Sr6 in *Triticum aestivum*. Can J Genet Cytol 23:183–190

Knott DR (1982) Multigenic inheritance of stem rust resistance in Wheat. Crop Sci 22:393–399

Knott DR (1984) The inheritance of resistance to race 56 of stem rust in 'Marquillo' wheat. Can J Genet Cytol 26:174–176

Knott DR, Dvorak J (1976) Alien germplasm as a source of resistance to disease. Annu Rev Phytopathol 14:211–235

Knott DR, McIntosh RA (1978) Inheritance of stem rust resistance in Webster wheat. Crop Sci 18:365–369

Kochman JK, Brown JF (1975) Host and environmental effects on post-penetration development of *Puccinia graminis avenae* and *P. coronata avenae*. Ann Appl Biol 81:33–41

Kochumadhavan M, Tomar SMS, Nambisan PNN (1980) Sources of rust resistance in wheat. Ind J Genet Plant Breed 40:610–618

Kogan M (1975) Plant resistance in pest management. In: Introduction to insects pest management. Wiley, New York, pp 103–146

Kogan M, Ortman EE (1978) Antixenosis – a new term proposed to replace Painters "Non-preference" modality of resistance. Entomol Soc Am Bull 24:175–176

Konzak CF (1956) Induction of mutations for disease resistance in cereals. In: Genetics in plant breeding. Brookhaven Symp Biol 9:141–156

Koshiary MA, Pan CL, Hak GE, Zaid ISA, Azizi A, Hindi C, Masoud M (1957) A study on the resistance of rice to stem borer infestations. Int Rice Commun Newsl 6:23–25

Kranz J (1974) Epidemiology, concepts and scope. In: Raychaudhuri SP, Verma JP (eds) Current trends in plant pathology. Lucknow Univ Dep Bot, Lucknow, pp 22–32

Krishna T, Seshu DV, Kalode MB (1977) New sources of resistance to brown planthopper of rice. Ind J Genet Plant Breed 37:147–153

Krishnamurthi M, Thaskal J (1974) Fiji disease resistant *Saccharum officinarum* var. Pindar subclones from tissue cultures. Proc Int Soc Sugar Cane Technol 15:130–137

Krull CF, Reyes ORJ, Elkin B (1965) Importance of "Small-uredia" reaction as an index of partial resistance to oat stem rust in Columbia. Crop Sci 6:494–497

Kubota T, Matsuura T (1953) Chemical studies on the black rot disease of sweet potato. J Chem Soc Jpn (Pure Chem Sect) 74:248–251

Kuc JA (1976) Phytoalexins. In: Heitefuss R, Williams PH (eds) Encyclopedia of plant physiology, vol IV. Springer, Berlin Heidelberg New York, pp 632–652

Kuc JA, Currier WW, Shih MJ (1976) Terpenoid phytoalexins. In: Friend J, Threlfall DR (eds) Biochemical aspects of plant – parasite relationships. Academic Press, London New York, pp 225–237

Kuhn RC, Ohm HW, Shaner GE (1978) Slow leaf rust severity in wheat against twenty two isolates of *Puccinia recondita*. Phytopathology 68:651–656

Kuhn RC, Ohm HW, Shaner GE (1980) Inheritance of slowleaf rusting resistance in Suwon 85 wheat. Crop Sci 20:655–659

Kulkarni RN, Chopra VL, Singh D (1982) Relative importance of components affecting the leaf rust progress curve in wheat. Theor Appl Genet 62:205–207

Kulshreshtha VP, Rao MV (1976) Genetics of resistance to an isolate of *Alternaria triticina* causing leaf blight in wheat. Euphytica 25:769–775

Kumar S, Singh RB (1981) Genetic analysis of adult plant resistance to powdery mildew in pea (*Pisum sativum* L.). Euphytica 30:147–151

Kumar S, Singh RM, Singh RB (1982) Inheritance of slow development of brown rust in wheat. Ind J Agric Sci 52:324–330

Kumar S, Shukla RS, Singh KP, Paxton JD, Husain A (1984) Glyceollin: A pathotoxin in leaf blight of *Costus speciosus*. Phytopathology 74:1349–1352

Kuo MO, Yoder C, Scheffer RP (1970) Comparative specificity of the toxins of *Helminthosporium carbonum* and *Helminthosporium victoriae*. Phytopathology 60:365–368

Kwon SH, Oh JH, Song HS, Lee JH (1974) Variation in pathogenicity of rice blast fungus induced by X-rays. SABRAO J 6:193–199

Lakshminarayana A, Khush GS (1977) New genes for resistance to the brown planthopper in rice. Crop Sci 17:96–100

Lal S, Singh IS (1984) Breeding for resistance to downy mildews and stalk rots in maize. Theor Appl Genet 69:111–119

Lamb RJ (1980) Hairs protect pods of mustard (*Brassica hirsuta* Gisilba) from flea beetle feeding damage. Can J Plant Sci 60:1439–1440

Lammerink J (1970) Inter-specific transfer of club root resistance from *Brassica campestris* L. to *B. napus* L. NZJ Agric Res 13:105–110

Lapwood DH (1961) Potato haulm resistance to *Phytophthora infestans*. I. Field assessment of resistance. Ann Appl Biol 49:140–151

Larson RI, Atkinson TG (1973) Wheat-*Agropyron* chromosome substitution lines as a source of resistance to wheat streak mosaic virus and its vector *Aceria* teelipae. In: Proc Symp 4th Int Wheat Genet, Columbia, pp 173–177

Larson RI, Atkinson TG (1981) Reaction of wheat to common root rot: Identification of a major gene, Crr, on chromosome 5B. Can J Genet Cytol 23:173–182

Larson RI, Atkinson TG (1982) Reaction of wheat to common root rot: Linkage of a major gene, Crr, with the centromere of chromosome 5B. Can J Genet Cytol 24:19–25

Laughnan JR, Gabay SJ (1973) Reaction of germinating maize pollen to *Helminthosporium maydis* pathotoxins. Crop Sci 13:681–684

Leath KT, Rowell JB (1970) Nutritional and inhibitory factors in the resistance of *Zea mays* to *Puccinia graminis*. Phytopathology 60:1097–1100

Leath S, Carroll RB (1982) Screening for resistance to *Fusarium oxysporum* in soybean. Plant Dis 66:1140–1143

Lebsock KL, Koch EJ (1968) Variation of stem solidness in wheat. Crop Sci 8:225–229

Lee JA (1971) Some problems in breeding smooth leaved cottons. Crop Sci 11:448–450

Legg PD, Litton CC, Collins GB (1980) Inheritance of resistance to the tobacco vein-mottling virus in tobacco. Can J Genet Cytol 22:21–26

Leisle D (1974) Genetics of leaf pubescence in durum wheat. Crop Sci 14:173–174

Lelley J (1957) Rostinfizierungsmethode im Dienste der pathologischen Resistenzzüchtung. Züchter 27:81–85

Leonard KJ (1969a) Factors affecting rates of stem rust increase in mixed plantings of susceptible and resistant oat varieties. Phytopathology 59:1845–1850

Leonard KJ (1969b) Selection in heterogeneous populations of *Puccinia graminis* f.sp. *avenae*. Phytopathology 59:1851–1857

Leonard KJ (1969c) Genetic equilibria in host-pathogen systems. Phytopathology 59:1858–1863

Leonard KJ (1977a) Selection pressures and plant pathogens. Ann NY Acad Sci 287:207–222

Leonard KJ (1977b) Virulence, temperature optima, and competitive abilities of isolines of race T and O of *Bipolaris maydis*. Phytopathology 67:1273–1279

Leonard KJ (1978) Polymorphisms for lesion type, fungicide tolerance, and mating capacity in *Cochliobolus carbonum* isolates pathogenic to corn. Can J Bot 56:1809–1815

Leonard KJ (1984) Population genetics of gene-for-gene interactions between plant host resistance and pathogen virulence. In: Chopra VL, Joshi BC, Sharma RP, Bansal BC (eds) Genetics, new frontiers, Proc XVth Int Congr Genet. IBH, Oxford, pp 131–148

Leonard KJ, Czochor RJ (1980) Theory of genetic interactions among populations of plants and their pathogens. Annu Rev Phytopathol 18:237–258

Leppik EE (1966) Searching gene centres of the genus *Cucumis* through host parasite relationship. Euphytica 15:323–338

Leppik EE (1968) Relation of centres of origin of cultivated plants to sources of disease resistance. US Dep Agric Plant Introd Invest Pap 13:1–8

Leppik EE (1970) Gene centres of plants as sources of disease resistance. Annu Rev Phytopathol 8:323–344

Letcher RM, Widdowson DA, Deverall BJ, Mansfield JW (1970) Identification and activity of wyerone acid as a phytoalexin in broad bean *(Vicia faba)* after infection by *Botrytis*. Phytochemistry 9:249–252

Librojo V, Kauffman HE, Khush GS (1976) Genetic analysis of bacterial blight resistance in four varieties of rice. SABRAO J 8:105–110

Lim SM (1975a) Diallel analysis for reaction of eight corn inbreds to *Helminthosporium maydis* race T. Phytopathology 65:10–15

Lim SM (1975b) Heterotic effects of resistance in maize to *Helminthosporium maydis* race O. Phytopathology 65:1117–1120

Lim SM, Hooker AL (1971) Southern corn leaf blight: genetic control of pathogenicity and toxin production in race T and race 0 of *Cochliobolus heterostrophus*. Genetics 69:115–117

Lim SM, Hooker AL (1972a) A preliminary characterization of *Helminthosporium maydis* toxins. Plant Dis Rep 56:805–807

Lim SM, Hooker AL (1972b) Disease deterimant of *Helminthosporium maydis* race T. Phytopathology 62:968–971

Lim SM, Paxton JD, Hooker AL (1968) Phytoalexin production in resistant to *Helminthosporium turcicum*. Phytopathology 58:720–721

Lim SM, Hooker AL, Paxton JD (1970) Isolation of phytoalexins from corn with monogenic resistance to *Helminthosporium turcicum*. Phytopathology 60:1071–1075

Lim SM, Hooker AL, Smith DR (1971) Use of *Helminthosporium maydis* race T pathotoxin to determine disease reaction of germinating corn seed. Agron J 63:712–713

Lim SM, Kinsey JG, Hooker AL (1974) Inheritance of virulence in *Helminthosporium turcicum* to monogenic resistant corn. Phytopathology 64:1150–1151

Lin KM, Lin PC (1960) Radiation induction variation in blast disease resistance in rice. Jpn J Breed 10:19–22

Lincolin RE, Cumins GB (1949) *Septoria* blight resistance in tomato. Phytopathology 39:647–655

Lincoln C, Dean G, Waddle BA, Yearian WC, Phillips JR, Roberts L (1971) Resistance of Frego-type cotton to boll weevil and boll worm. J Econ Entomol 64:1326–1327

Lindberg GD (1971) Disease induced toxin produced in *Helminthosporium oryzae*. Phytopathology 61:420–424

Lindley G (1831) A guide to the orchard and kitchen garden. In: I. Lindley (ed) London, Longmans, Rees, Orme, Brown, and Green, p 601

Ling KC (1972) Rice virus diseases. Int Rice Res Inst, Philippines, pp 142

Ling KC (1974) An improved mass screening method for testing the resistance of rice varieties to tungro disease in the green house. Philipp Phytopathol 10:19–30

Ling KC, Aquiero VM, Lee SH (1970) A mass screening method for testing resistance to grassy stunt disease of rice. Plant Dis Rep 54:565–569

Little R (1971) An attempt to induce resistance to *Septoria nodorum* and *Puccinia graminis* in wheat using gamma rays, neutrons and EMS as mutagenic agents. In: Proc Symp Mutat Breed Dis Resist. Int At Energy Ag, Vienna, pp 139–149

Littlefield LJ (1969) Flax rust resistance induced by prior inoculation with an avirulent race of *Melampsora lini*. Phytopathology 59:1323–1328

Livers RW, Harvey TL (1969) Green bug resistance in rye. J Econ Entomol 62:1368–1370

Loegering WQ, Browder LE (1971) A system of nomenclature for physiologic races of *Puccinia recondita tritici*. Plant Dis Rep 55:718–722

Loegering WQ, Powers HR Jr (1962) Inheritance of pathogenicity in a cross of physiological races 111 and 36 of *Puccinia graminis* f.sp. *tritici*. Phytopathology 52:547–554

Loegering WQ, Sleper DA, Johnson JM, Asay KH (1976) Rating general resistance on a single-plant basis. Phytopathology 66:1445–1448

Long BJ, Dunn GM, Bowman JS, Routley DG (1976) Relation of hydroxamic acid concentration (DIMBOA) to resistance to the corn leaf aphid. Maize Genet Co-op Newsl 50:91

Long BJ, Dunn GM, Bowman JS, Routley DG (1977) Relationship of hydroxamic acid content in corn and resistance to the corn leaf aphid. Crop Sci 17:55–58

Loomis RS, Beck SD, Stauffer JE (1957) The European corn borer, *Pyrausta nubilalis* (Hubn.) and its principal host plant. V. A chemical study of host plant resistance. Plant Physiol 32:379–385

Lorz H, Scowcroft WR (1983) Variability among plants and their progeny regenerated from protoplasts of Su/su heterozygotes of *Nicotiana tabaccum*. Theor Appl Genet 66:67–75

Lowe HJB (1984) Characteristics of resistance to grain aphid *Sitobion avenae* in winter wheat. Ann Appl Biol 105:529–538

Luckmann WH, Rhodes AM, Wann EV (1964) Silk balling and other factors associated with resistance of corn to corn ear worm. J Econ Entomol 57:778–779

Luginbill P Jr (1969) Developing resistant plants – the ideal method of controlling insects. US Dep Agric ARS Prod Rep 111:14

Luke HH, Murphy HC, Petr FC (1966) Inheritance of spontaneus mutations of the Victoria locus in oats. Phytopathology 56:210–212

Luke HH, Chapman WH, Barnett RD (1972) Horizontal resistance of Red Rust Proof Oats to crown rust. Phytopathology 62:414–417

Lukefahr MJ, Houghtaling JE (1969) Resistance of cotton strains with high gossypol content to *Heliothis* spp. J Econ Entomol 62:588–591

Lukefahr MJ, Martin DF (1966) Cotton plant pigments as a source of resistance to the boll worm and tobacco bud worm. J Econ Entomol 59:176–179

Lukefahr MJ, Noble LW, Houghtaling JE (1966) Growth and infestation of bollworms and other insects on glanded and glandless strains of cotton. J Econ Entomol 59:817–820

Lukefahr MJ, Houghtaling JE, Graham HM (1971) Suppression of *Heliothis* populations with glabrous cotton strains. J Econ Entomol 64:486–488

Luthra JK, Chandra S (1983) Inheritance of resistance to hill bunt in wheat. Ind J Genet Plant Breed 43:318–320

Luthra JK, Rao MV (1973) Inheritance of seedling resistance to *Helminthosporium* blight in barley. Ind J Genet Plant Breed 33:378–379

Luthra JK, Rao MV (1979) Multiline cultivars – How their resistance influence leaf rust diseases in wheat. Euphytica 28:137–144

Luthra JK, Rao MV (1980) Concepts and strategies to control leaf rust epidemic of wheat in India. Ind J Genet Plant Breed 40:180–186

Lyman JM, Cardona C (1982) Resistance in Lima beans to a leafhopper *Empoasca kraemeri*. J Econ Entomol 75:281–286

Macor RF (1960) Nature and exploitation of crop plant resistance to disease. Nature (London) 186:857

MacKenzie DR (1976) Application of two epidemiological models for identification of slow rusting in wheat. Phytopathology 66:55–59

MacKenzie DR (1978) Estimating parasite fitness. Phytopathology 68:9–13

MacKey J (1977) Strategies of race – specific phytoparasitism and its control by plant breeding. Genetika 9:237–255

Maclean DJ, Sargent JA, Tommerup IC, Ingram DS (1974) Hypersensitivity as the primary event in resistance to fungal parasites. Nature (London) 249:186–187

Mahadevan A, Kuc J, Williams EB (1965) Biochemistry of resistance in cucumber against *Cladosporium cucumerinum*. I. Presence of a pectinase inhibitor in resistant plants. Phytopathology 55:1000–1003

Mai WF, Peterson LC (1952) Resistance of *Solanum balsii* and *S. sucrense* to the golden nematode, *Heterodera rostochiensis* Woll. Science 116:224

Main CE (1968) Induced resistance to bacterial wilt in susceptible tobacco cuttings pretreated with avirulent mutants of *Pseudomonas solanacearum*. Phytopathology 58:1058–1059 (Abstr)

Malcolmson JF (1970) Vegetative hybridity in *Phytophthora infestans*. Nature (London) 225:971–972

Maltais JB, Auclair JL (1957) Factors in resistance of peas to the pea aphid, *Acyrthosiphon pisum* (Harr.) *(Homoptera:Aphididae)* I. The sugar nitrogen ratio. Can Entomol 89:365–370

Manglitz GR, Gorz HJ (1968) Inheritance of resistance in sweet clover to the sweet clover aphid. J Econ Entomol 61:90–93

Marke ME, Meyer JR (1963) Studies of resistance to cotton strains to the boll weevil. J Econ Entomol 56:860–862

Marshall DR (1977) The advantages and hazards of genetic homogeneity. Ann NY Acad Sci 287: 1–20

Marshall DR, Pryor AJ (1978) Multiline varieties and disease control. I. The dirty crop approach with each component carrying a unique single resistance gene. Theor Appl Genet 51:177–184

Marshall DR, Pryor AJ (1979) Multiline varieties and disease control. II. Dirty crop approach with components carrying 2 or more genes for resistance. Euphytica 28:145–159

Marston AR (1930) Breeding corn for resistance to European corn borer. J Am Soc Agron 22: 986–992

Martens JW, McKenzie RIH, Fleischmann G (1968) The inheritance of resistance to stem and crown rust in Kyoto Oats. Can J Genet Cytol 10:808–812

Martens JW, McKenzie RIH, Green GJ (1970) Gene-for-gene relationships in the *Avena*: *Puccinia graminis* host-parasite system in Canada. Can J Bot 48:969–975

Martens JW, Rothman PG, McKenzie RIH, Brown PD (1981) Evidence of complementary gene action confering resistance *Puccinia graminis avenae* in *Avena sativa*. Can J Genet Cytol 23:591–595

Martin JT (1964) Role of cuticle in the defense against plant disease. Annu Rev Phytopathol 2: 81–100

Martin MV (1962) *Solanum pennelli*, A possible source of tomato curly top resistance. Phytopathology 52:1230

Martin MV, Thomas PE (1969) A new tomato line resistant to curly top virus. Phytopathology 59: 1754–1755

Martinez CR, Khush GS (1974) Sources and inheritance of resistance to brown plant hopper in some breeding lines of rice. Crop Sci 14:264–267

Mastenbrock C (1955) A note on resistance of *Solanum* species to powdery mildew. Euphytica 4: 15–16

Matern U, Strobel G, Shepard J (1978) Reaction to phytotoxins in a potato population derived from mesophyll protoplasts. Proc Natl Acad Sci USA 75:4935–4937

Maxwell FG, Parrott WL, Jenkins JN, Lafever HN (1965) A boll weevil feeding deterrent from the calyx of an alternate host, *Hibiscus syriacus*. J Econ Entomol 58:985–988

Maxwell FG, Jenkins JN, Parrott WL, Buford WT (1969) Factors contributing to resistance and susceptibility of cotton and other pests of the boll weevil, *Anthonomus grandis*. Entomol Exp Appl 12:801–810

Maxwell FG, Jenkins JN, Parrott WL (1972) Resistance of plants to insects. Adv Agron 24:187–265

Mayama S, Daly JM, Rehfeld DW, Daly CR (1975) Hypersensitive response of near isogenic wheat carrying the temperature – sensitive Sr6 allele for resistance to stem rust. Physiol Plant Pathol 7:35–47

McFarlane JS, Hartzler E, Frazier WA (1946) Breeding tomatoes for nematode resistance and for high vitamin C content in Hawaii. Proc Am Soc Hortic Sci 47:262–270

McKenzie H (1965) Inheritance of sawfly reaction and stem solidness in spring wheat crosses. Can J Plant Sci 45:583–589

McMurty JA, Stanford EH (1960) Observations of feeding habits of the spotted alfalfa aphid on resistant and susceptible alfalfa plants. J Econ Entomol 53:714–717

McNeal FH, Lebsock KL, Luginbill P Jr, Noble WB (1957) Segregation for stem solidness in crosses of Rescue and 4 Portuguese Wheats. Agron J 49:246–248

McNeal FH, Wallace LE, Berg MA, McGuire CF (1971) Wheat stem sawfly resistance with agronomic and quality information in hybrid wheats. J Econ Entomol 64:939–941

McNew GL (1972) Concept of pest management. In: Pest control strategies for the future. Natl Acad Sci, Washington DC, pp 119–133

Melouk HA, Sanborn MR, Banks DJ (1984) Sources of resistance to peanut mottle virus in *Arachis* germplasm. Plant Dis 68:563–564

Merkle OG, Smith EL (1983) Inheritance of resistance to soil born mosaic in wheat. Crop Sci 23: 1075–1076

Metcalf RL, Rhodes AM, Metcalf RA, Ferguson J, Metcalf ER, Lu Po-Yung (1982) Cucurbitacin contents and Diabroticite (Coleoptera:Chrysomelidae) feeding upon *Cucurbita* spp. Environ Entomol 11:931–937

Metlitskii LV, Ozeretskovskaya OL, Vul'fgon NS, Chalova LI (1971) Chemical nature of lubinin, a new phytoalexin of potatoes. Dokl Akad Nauk USSR 200:1470–1472

Mettler LE, Gregg TG (1969) Population genetics and evolution. Prentice-Hall, Englewood Cliffs, NJ, p 212

Metzger RJ, Trione EJ (1962) Application of the gene-for-gene relationship hypothesis to the *Triticum Tilletia* system. Phytopathology 52:363 (Abstr)

Metzger RJ, Schaller CW, Rohde CR (1979) Inheritance of resistance to common bunt in wheat, C.I., 7090. Crop Sci 19:309–312

Mew TW, Khush GS (1981) Breeding for bacterial blight resistance in rice. Proc 5th Int Conf Plant Pathol Bacteriol Calif, Colombia, CIAT, pp 504–510

Mew TW, Wu SZ, Horino O (1982) Pathotypes of *Xanthomonas campestris* pv. *oryzae* in Asia. IRRI Res Pap No 75. Int Rice Res Inst, Philippines, p 7

Meyer JR (1957) Origin and inheritance of D_2 smoothness in upland cotton. J Hered 48:249–250

Meyer JR, Meyer VG (1961) Origin and inheritance of nectariless cotton. Crop Sci 1:167–169

Miah MAJ (1968) Method of selfing F_1 cultures and interpreting genetic data in rust fungi. Can J Genet Cytol 10:613–619

Micke A (1983) Some considerations on the use of induced mutations for improving disease resistance of crop plants. In: Proc Symp Induced Mutat Dis Resist Crop Plants. Int At Energy Ag, Vienna, pp 3–19

Micke A, Hsieh SC, Sigurbijornsson B (1971) Rice breeding with induced mutations. In: Proc Symp Int Rice Res Inst, Philippines, pp 573–580

Miles JW, Dudley JW, White DG, Lambert RJ (1980) Improving corn population for grain yield and resistance to leaf blight and stalk rot. Crop Sci 20:247–251

Miller BS, Robinson RJ, Johnson JA, Jones ET, Ponnaiya BWX (1960) Studies on the relation between silica in wheat plants and resistance to hessianfly attack. J Econ Entomol 53:995–999

Modawi RS, Heyne EG, Brunetta D, Wills WG (1982) Genetic studies of field reaction to wheat soilborne mosaic virus. Plant Dis 66:1183–1184

Mode CJ (1958) A mathematical model for the co-evolution of obligate parasites and their hosts. Evolution 12:158–165

Mode CJ (1961) A generalized model of a host pathogen system. Biometrics 17:386–404

Moore FJ (1977) Disease epidemics and host genetic resistance in British crops. Ann NY Acad Sci 287:21–28

Mortensen JA (1981) Sources and inheritance of resistance to anthracnose in *Vitis*. J Hered 72: 423–426

Moseman JG (1957) Host parasite interactions between culture 12 A1 of the powdery mildew fungus and the Mlk and Mlg genes in barley. Phytopathology 47:453 (Abstr)

Moseman JG (1959) Host-pathogen interaction of the genes for resistance in *Hordeum vulgare* and for pathogenicity in *Erysiphe graminis* f.sp. *hordei*. Phytopathology 49:469–472

Moseman JG, Greeley LW (1966) Reaction of one barley plant to several cultures of *Erysiphe graminis* f.sp. *hordei*. Phytopathology 56:1428–1429

Moseman JG, Jorgensen JH (1973) Differentiation of resistance genes at the Ml-a locus in six pairs of isogenic barley lines. Euphytica 22:189–196

Moseman JG, Baenziger PS, Kilpatrick RA (1981) Genes conditioning resistance of *Hordeum spontaneum* to *Erysiphe graminis* f.sp. *hordei*. Crop Sci 21:229–232

Muko H, Yoshida K (1951) A needle inoculation method for bacterial leaf blight disease of rice. Ann Phytopathol Soc Jpn 15:179

Müller KO (1958) Studies on phytoalexins. I. The formation and immunological significance of phytoalexin produced by *Phaseolus vulgaris* in response to infections with *Sclerotina fructicola* and *Phytophthora infestans*. Aust J Biol Sci 11:275–300

Müller KO, Börger H (1941) Experimentelle Untersuchungen über die *Phytophthora* Resistenz der Kartoffel. Zugleich ein Beitrag zum Problem der erworbenen Resistenz im Pflanzenreich. Arb Biol Reichanst Berlin 23:189–231

Mumford D (1972) A new method of mechanically transmitting curly top virus. Phytopathology 62:1217

Munakata K, Okamoto D (1967) Varietal resistance to rice stem borers in Japan. In: Major insect pests of the rice plant. Hopkins, Baltimore, pp 419–430

Murray MJ (1961) Spearmint rust resistance and immunity in genus *Mentha*. Crop Sci 1:175–179

Murray MJ (1969) Successful use of irradiation breeding to obtain *Verticillium* resistant strains of peppermint, *Mentha piperita* L. In: Proc Symp Induced Mutat Plants. Int At Energy Ag, Vienna, pp 345–371

Murray MJ (1971) Additional observations on mutation breeding to obtain *Verticillium* – resistant strains of peppermint. In: Proc Symp Mutat Breed Dis Resist. Int At Energy Ag, Vienna, pp 171–195

Muttuthamby S, Aslam M, Khan MA (1969) Inheritance of leaf hairiness in *Gossypium hirsutum* L. cotton and its relationship with jassid resistance. Euphytica 18:435–439

Nabors MV (1976) Using spontaneously occurring and induced mutations to obtain agriculturally useful plants. Bioscience 26:761–767

Nagaich BB, Upadhya MD, Prakash O, Singh SJ (1968) Cytoplasmically determined expression of symptoms of potato virus X in crosses between species of *Capsicum*. Nature (London) 220:1341–1342

Nair RV, Masajo TM, Khush GS (1982) Genetic analysis of resistance to white backed planthopper in twenty-one varieties of rice, *Oryza sativa* L. Theor Appl Genet 61:19–22

Nakano K, Abe G, Taketa N, Hirano C (1961) Silicon as an insect resistance component of host plant, found in the relation between the rice stem borer and rice plant. Jpn J Appl Entomol Zool 5:17–21

Nanda JS, Singh DP, Chaudhary RC (1977) Bacterial leaf streak disease of rice. IL RISO XXVI (3): 235–241

Narayana D, Prasad MN (1983) Inheritance of resistance to *Fusarium* grain mould in sorghum. 15th Inst Congr Genet Abstr No 1316:727

Narula PN, Luthra JK, Chandra S (1983) Inheritance of seedling resistance to an isolate of leaf rust in barley. Ind J Genet Plant Breed 43:454–456

Narula PN, Srivastava OP, Prasad RS (1984) Genetics of resistance to corn leaf aphid in barley. Ind J Genet Plant Breed 44:177–180

Nath P, Hall CV (1963) Inheritance of resistance to the striped cucumber beetle in *Cucurbita pepo*. Ind J Genet Plant Breed 23:342–345

Nath P, Dutta OP, Sundri V (1976a) Breeding pumpkin for resistance to fruitfly. SABRAO J 8:29–33

Nath P, Dutta OP, Sundari V, Swamy KRM (1976b) Inheritance of resistance to fruitfly in pumpkin. SABRAO J 8:117–119

Nathawat KS, Jain R, Gandhi SM (1979) New resistance genes in wheat to Indian races of stem rust. In: Proc Symp 5th Int Wheat Genet New Delhi 2:1031–1038

Nayar JK, Frankel G (1963) The chemical basis of host selection for the Mexican bean beetle, *Epilechna varivestris*. Ann Entomol Soc Am 56:174–178

Nelson LR, Gates CE (1982) Genetics of host plant resistance of wheat to *Septoria nodorum*. Crop Sci 22:771–773

Nelson RR (1972) Stabilizing racial populations of plant pathogens by use of resistance genes. J Environ Qual 1:220–227

Nelson RR (ed) (1973) The meaning of disease resistance in plants. In: Nelson RR (ed) Breeding plants for disease resistance: concepts and application. Pa State Univ Press, Univ Park London, pp 13–25

Nelson RR (1975) Horizontal resistance in plants: concepts, controversies and applications. In: Galvez GE (ed) Proc Semin on Horizontal Resist Blast Dis Rice. Ser CE-9 CIAT, Cali Colombia, pp 1–20

Nelson RR (1978) Genetics of horizontal resistance to plant diseases. Annu Rev Phytopathol 16:359–378

Nelson RR (1979) The evolution of paraiste fitness. In: Horsfall JG, Cowling EB (eds) Plant disease, vol IV. How pathogen induce disease. Academic Press, London New York, pp 23–45

Nelson RR, Kline DM (1969) Genes for pathogenicity in *Cochliobolus heterostrophus*. Can J Bot 47:1311–1314

Nelson RR, Wilcoxson RD, Christensen JJ (1955) Heterokaryosis as a basis for variation in *Puccinia graminis* var. *tritici*. Phytopathology 45:639–643

Nene YL (1972) A survey of the viral diseases of pulse crops in Uttar Pradesh. GB Plant Univ Agric Tech Pantnagar India Exp Stn Bull No 4:191

Nene YL, Reddy MV (1976) Screening for resistance to sterility mosaic of pigeonpea. Plant Dis Rep 60:1034–1036

Newton M, Johnson T, Brown AM (1930) A study of the inheritance of spore colour and pathogenicity in crosses between physiologic forms of *Puccinia graminis tritici*. Sci Agric 10:775–798

Nicolaisen W (1934) Die Grundlagen der Immunitätszüchtung gegen *Ustilago avenae* (Pers.). Jens Z Zuecht A Aflzuecht 19:1

Niederhauser JS (1962) Evaluation of multigenic "field resistance" of the potato to *Phytophthora infestans* in 10 years of trials at Toluca, Mexico. Phytopathology 52:746 (Abstr)

Niederhauser JS, Cervantes J (1956) Maintenance of field resistance to *Phytophthora infestans* in potato selection. Phytopathology 46:22 (Abstr)

Niederhauser JS, Mills WR (1953) Resistance of *Solanum* species to *Phytophthora infestans* in Mexico. Phytopathology 43:456–457

Niederhauser JS, Cervantes J, Servia L (1954) Late blight in Mexico and its implications. Phytopathology 54:406–408

Nielson MV, Don H, Schonhorst MH, Lehman WF, Marble VL (1970) Biotypes of the spotted alfalfa aphid in varieties in western United States. J Econ Entomol 63:1822–1925

Nielson MW, Kuehl RO (1982) Screening efficacy of spotted alfalfa aphid biotypes and genic systems for resistance in alfalfa. Environ Entomol 11:989–996

Nielson MW, Olson DL (1982) Horizontal resistance in 'Lahontan' alfalfa to biotypes of the spotted alfalfa aphid (Homoptera: Aphididae). Environ Entomol 11:928–930

Nilan RA, Kleinhofs A, Konzak CF (1977) The role of induced mutation in supplementing natural genetic variability. Ann NY Acad Sci 287:367–384

Nilsson-Ehle H (1920) Über die Resistenz gegen *Heterodera schachtii* bei gewissen Gerstensorten, ihre Vererbungsweise und Bedeutung für die Praxis. Hereditas 1:1–34

Nitsch JP (1972) Haploid plants from pollen. Z Pflanzenzuecht 67:3–18

Nitzsche W (1983) Chitinese as a possible resistance factor for higher plants. Theor Appl Genet 65:171–172

Nitzsche W, Wenzel W (1977) Haploids in plant breeding. Parey, Berlin, 101 pp

Nobel WB, Suneson CA (1943) Differentiation of two genetic factors for resistance to the hessianfly in Dawson Wheat. J Agric Res 67:27–32

Noronha-Wagner M, Bettencourt AJ (1967) Genetic study of the resistance of *Coffea* spp. to leaf rust. 1. Identification and behaviour of four factors conditioning disease reaction in *Coffea arabica* to twelve physiologic races of *Hemileia vastatrix*. Can J Bot 45:2021–2031

Norris DM, Kogan M (1979) Biochemical and morphological basis of resistance. In: Maxwell FG, Jennings PR (eds) Breeding plants resistant to insects. Wiley Interscience, pp 23–61

Nuque FL, Aguiero VM, Ou SH (1982) Inheritance of resistance to grassy stunt virus in rice. Plant Dis 66:63–64

Nutman PS (1969) Genetics of symbiosis and nitrogen fixation in legumes. Proc R Soc London Ser B 172:417–437

Nwigwe C (1973) Influence of epiphytic yellow bacteria on the pathogenicity of *Xanthomonas oryzae* on rice leaves. IL RISO XXII (2):191–192

Nyquist WE (1962) Differential fertilization in the inheritance of stem rust resistance in hybrids involving a common wheat strain derived from *Triticum timopheevi*. Genetics 47:1109–1124

O'Brien VJ, Leach SS (1983) Investigations into the mode of resistance of potato tubers to *Fusarium roseum* Sambucinum. Am Potato J 60:227–233

Odiemah M, Manninger I (1982) Inheritance of resistance to *Fusarium* ear rot in maize. Acta Phytopathol Acad Sci Hung 17:91–99

Oettermann CM, Patterson FL, Gallun RL (1983) Inheritance of resistance in 'Luso' wheat to hessian fly. Crop Sci 23:221–224

Ogle HJ, Brown JF (1971) Quantitative studies of the post penetration phase of infection by *Puccinia graminis tritici*. Ann Appl Biol 67:309–319

Ogle HJ, Taylor NW, Brown JF (1973) A mathematical approach to the production of differences in the relative ability of races of *Puccinia graminis tritici* to survive when mixed. Aust J Biol Sci 26:1137–1143

Ohm HW, Shaner GE (1976) Three components of slow leaf rusting of different growth stages in wheat. Phytopathology 66:1356–1360

Ohyama K, Nitsch JP (1972) Flowering haploid plants from protoplasts on tobacco leaves. Plant Cell Physiol 13:229–236

Olembo JR, Patterson FL, Gallun RL (1966) Genetic analysis of the resistance to *Mayetiola destructor* (Say) in *Hordeum vulgare* L. Crop Sci 6:563–566

Oliver BF, Gifford JR (1975) Weight differences among stalk borer larvae collected from rice lines showing resistance in field studies. J Econ Entomol 68:134

Oliver BF, Maxwell FG, Jenkins JN (1970) A comparison of the damage done by the bollworm to glanded and glandless cottons. J Econ Entomol 63:1328–1329

Oliver BF, Maxwell FG, Jenkins JN (1971) Growth of the bollworm on glanded and glandless cotton. J Econ Entomol 64:396–398

Onukogu FA, Guthrie WD, Russell WA, Reed GL, Robbins JC (1978) Location of genes that condition resistance in maize to sheath-collar feeding by second generation European corn borer. J Econ Entomol 71:1–4

Oort AJP (1963) A gene-for-gene relationship in the *Triticum Ustilago* system and some remarks on host-parasite combinations in general. Neth J Plant Pathol 69:104–109

Orellana RG (1981) Resistance to bud blight in introductions from the germplasm of wild soybean. Plant Dis 65:594–595

Orton TJ, Durgan ME, Hulbert SD (1984) Studies on the inheritance of resistance to *Fusarium oxysporum* f.sp. *apii* in Celery. Plant Dis 68:574–578

Orton WA (1900) The wilt disease of cotton and its control. USDA Div Veg Physiol Pathol Bull 27

Orton WA (1909) The development of farm crops resistant to disease. US Dep Agric Yearb 1908: 453–464

Osuna JA, Araujo SMC de, Lara FM (1983) Evaluation of two cycles of recurrent selection for corn earworm resistance. 15th Int Congr Genet Abstr No 1317:728

Ou SH (1979) Int Rice Res Inst, Philippines. Rice Blast Workshop

Ou SH, Nuque FL, Bandong JM (1975) Relation between qualitative and quantitative resistance to rice blast. Phytopathology 65:1315–1316

Padmanabhan SY, Kaur S, Rao M (1977) Induction of resistance to bacterial leaf blight *(Xanthomonas oryzae)* disease in the high yielding variety, Vijaya (IR 8 X T 90). In: Proc Symp Induced Mutat Against Plant Dis. Int At Energy Ag, Vienna, pp 187–198

Paharia KD, Pushkarnath, Deshmukh MJ (1962) Resistance to potato varieties to charcoal rot. Ind Potato J 4:84–87

Painter RH (1936) The food of insects and its relation to resistance of plants to insect attack. Am Nat 70:547–566

Painter RH (1941) The economic value and biological significance of plant resistance to insect attack. J Econ Entomol 34:358–367

Painter RH (1951) Insect resistance in crop plants. Macmillan, New York, p 520

Painter RH (1958) Resistance of plants to insects. Annu Rev Entomol 3:267–290

Painter RH, Pathak MD (1962) The distinguishing features and significance of the four biotypes of the corn leaf aphid, *Rhopalosiphum maidis* (Fitch). 11th Int Kongr Entomol, Wien, 1960. Sonderdr Verh II:110–115

Painter RH, Peters DC (1956) Screening wheat varieties and hybrids for resistance to the greenbug. J Econ Entomol 49:546–548

Pal AB, Srinivasan K, Doijode SD (1984) Sources of resistance to melon fruitfly in bittergourd and possible mechanisms of resistance. SABRAO J 16:57–64

Panda N (1979) Principles of Host-plant resistance to insect-pests. Hindustan, Delhi, p 386

Panda N, Pradhan B, Samalo AP, Prakasarao PS (1975) Note on the relationship of some biochemical factors with the resistance in rice varieties to yellow rice borer. Ind J Agric Sci 45:499–501

Pandey KC, Singh A, Faruqui SA (1984) Sources of resistance to spotted alfalfa aphid (*Therioaphis maculata* Buckton) in Medics. Ind J Genet Plant Breed 44:1–6

Panwar VPS, Siddiqui KH, Marwaha KK, Sarup P (1979) Location of sources of resistance to the maize stalk borer, *Chilo partellus* (Swinhoe) a most introductory nursery comprising exotic and indigenous maize germplasm. J Entomol Res 3:191–195

Parlevliet JE (1976) Evaluation of the concept of horizontal resistance in the barley *Puccinia hordei* host pathogen relationship. Phytopathology 66:494–497

Parlevliet JE (1977) Evidence of differential interaction in the polygenic *Hordeum vulgare–Puccinia hordei* relation during epidemic development. Phytopathology 67:776–778

Parlevliet JE (1978) Race-specific aspects of polygenic resistance of barley to leaf rust, *Puccinia hordei*. Neth J Plant Pathol 84:121–126

Parlevliet JE (1981) Disease resistance in plants and its consequences for plant breeding. In: Frey KJ (ed) Plant breeding. Iowa State Univ Press, Ames USA, pp 309–364

Parlevliet JE (1983) Can horizontal resistance be recognized in the presence of vertical resistance in plants exposed to a mixture of pathogen races? Phytopathology 73:379

Parlevliet JE, Kuiper HJ (1977) Partial resistance of barley to leaf rust, *Puccinia hordei*. IV. Effect of cultivar and development stage on infection frequency. Euphytica 26:249–255

Parlevliet JE, Ommeren AV (1975) Partial resistance of barley to leaf rust, *Puccinia hordei*. II. Relationship between yield trials, micro plot tests and latent period. Euphytica 24:293–303

Parlevliet JE, Zadoks JC (1977) The integrated concept of disease resistance: A new view including horizontal and vertical resistance in plants. Euphytica 26:5–21

Parlevliet JE, Lindhout WH, Van Ommeren A, Kuiper HJ (1980) Level of partial resistance to leaf rust, *Puccinia hordei* in West-European barley and how to select for it. Euphytica 29:1–8

Parnell FR, King HE, Ruston DF (1949) Jassid resistance and hairiness of the cotton plant. Bull Entomol Res 39:539–572

Patanakmjorn S, Pathak MD (1967) Varietal resistance of the Asiatic rice borer, *Chilo suppressalis* (Lepidoptera:Crambide) and its association with various plant characteristics. Ann Entomol Soc Am 60:287–292

Patel PN (1978) Bacterial pustule disease in Tanzania: Pathogenic variability and host resistance. In: Abstr Pap 3rd Int Congr Plant Pathol, Munich, p 78

Patel PN (1983) Resistance to bacterial blight in cowpeas in Tanzania and other countries. Ind J Genet Plant Breed 43:9–13

Patel PN, Walker JC (1963) Changes in free aminoacids and amide content of resistant and susceptible beans after infection with the halo blight organism. Phytopathology 53:522–528

Patel PN, Mligo JK, Leyna HK, Kuwite C, Mmbaga ET (1982a) Sources of resistance and breeding of cowpeas for resistance to a strain of cowpea aphid borne mosaic virus from Tanzania. Ind J Genet Plant Breed 42:221–229

Patel PN, Mligo JK, Leyna HK, Kuwite C, Mmbaga ET (1982b) Multiple disease resistance cowpea breeding program in Tanzania. Ind J Genet Plant Breed 42:230–239

Pathak MD (1961) Preliminary notes on the differential response of yellow and brown sarson and rai to mustard aphid, *Lipaphis erysimi* (Kalt.). Ind Oilseeds J 5:39–44

Pathak MD (1967) Varietal resistance to rice borers at IRRI. In: Proc Symp Major Insect Pests Rice Plant. Hopkins, Baltimore, pp 405–418

Pathak MD (1969) Stem borer and leafhopper-planthopper resistance in rice varieties. Entomol Exp Appl 12:789–800

Pathak MD (1970) Genetics of plants in pest management. In: Rabb RL, Guthrie FENC (eds) Concepts of pest management. State Univ, Raleigh, pp 138–157

Pathak MD (1971) Resistance to insect-pests in rice varieties. *Oryza* (Suppl) 8(2):135–144

Pathak MD (1975) Utilization of insect-pest interrelationships in pest control. In: Pimental D (ed) Insects, science and society. Academic Press, London New York, pp 121–148

Pathak MD (1977) Defense of the rice crop against insect pests. In: Day PR (ed) The genetic basis of epidemics in agriculture. Ann NY Acad Sci 287:287–295

Pathak MD, Saxena RC (1980) Breeding approaches in rice. In: Maxwell FG, Jennings PR (eds) Breeding plants resistance to insects. Wiley Interscience, New York, pp 422–454

Pathak MD, Chang CH, Fortuno HE (1969) Resistance of *Nilaparvata lugens* in varieties of rice. Nature (London) 223:502–504

Patterson FL, Gallun RL (1973) Inheritance of Seneca wheat to race E of hessian-fly. In: Proc 4th Int Wheat Genet Symp Mo Agric Exp Stn, Columbia, pp 445–449

Patterson FL, Gallun RL (1977) Linkage in wheat of the H_3 and H_6 genetic factors for resistance to hessian-fly. J Hered 68:293–296

Paxton JD, Chamberlain DW (1969) Phytoalexin production and disease resistance in soybeans as affected by age. Phytopathology 59:775–777

Paxton RN, Hunt OJ, Faulkner LR, Griffin GD, Jensen HJ, Stanford EH (1976) Registration of a multiple-pest resistant alfalfa germplasm. Crop Sci 16:125–126

Peaden RN, Hunt OJ, Faulkner LR, Griffin GD, Jensen HJ, Stanford EH (1976) Registration of multiple pestresistant alfalfa germplasm. Crop Sci 16:125–126

Pedersen MV, Barnes DK, Sorensen EL, Griffin GD, Nielson MV, Hill RR, Frosheiser FI, Sonoda RM, Hanson CH, Hunt DJ, Peadon RN, Elgin JH, Devine TE, Anderson MJ, Goplen BP, Elling LJ, Howarth RE (1976) Effects of low and high saponin selection in alfalfa on agronomic and pest resistance traits and interrelationship of these traits. Crop Sci 16:193–199

Pelcher LE, Kao KN, Gamborg OL, Yodder OC, Gracen VE (1975) Effects of *Helminthosporium maydis* race T toxin on protoplasts of resistant and susceptible corn *(Zea mays)*. Can J Bot 53: 427–431

Pelham J (1966) Resistance in tomato to tobacco mosaic virus. Euphytica 15:258–267

Penny LH, Dicke FF (1956) Inheritance of resistance in corn to leaf feeding by the European corn borer. Agron J 48:200–203

Penny LH, Dicke FF (1957) A single gene-pair controlling segregation for European corn borer resistance. Agron J 49:193–196

Penny LH, Scott GE, Guthrie WD (1967) Recurrent selection of European corn borer resistance in maize. Crop Sci 7:407–409

Perrin DR (1964) The structure of phaseollin. Tetrahedron Lett 1964:29–35

Perrin DR, Bottomley W (1962) Studies on phytoalexins. V. The structure of pisatin from *Pisum sativum* L. J Am Chem Soc 84:1919–1922

Perrin DR, Whittle CP, Batterham TJ (1972) The structure of phaseollidin. Tetrahedron Lett 1972: 1673–1676

Person C (1959) Gene-for-gene relationships in host: Parasite systems. Can J Bot 37:1101–1130

Person C, Ebba T (1975) Genetics of fungal pathogens. Genetics (Suppl) 79:397

Person C, Mayo GME (1974) Genetic limitations of models of specific interactions between a host and its parasite. Can J Bot 52:1339–1347

Person C, Sidhu G (1971) Genetics of host-parasite interrelationships. In: Proc Symp Mutat Breed Dis Resist. Int At Energy Ag, Vienna, pp 31–38

Peterson RF, Campbell AB (1953) Aneuploid analyses of the genes for stem rust resistance and head density in McMurachy Wheat. Rep Int Wheat-Stem-Rust Conf, Winnipeg (Mimeo), pp 99–100

Petpisit V, Khush GS, Kauffman HE (1977) Inheritance of resistance to bacterial blight in rice. Crop Sci 17:551–554

Phillips MS, Dale MFB (1982) Assessing potato seedling progenies for resistance to the white potato cyst nematode. J Agric Sci 99:67–70

Pitt D, Coombes C (1969) Release of hydrolytic enzymes from cytoplasmic particles of *Solanum* tuber tissues during infection by tuber-rotting fungi. J Gen Microbiol 56:321–329

Platt AW, Darrock JG, Kemp HJ (1941) The inheritance of solid stem and certain other characteristics in crosses between varieties of *Triticum vulgare*. Sci Agric 22:216–223

Pontecorvo G (1956) The parasexual cycle in fungi. Annu Rev Microbiol 10:393–400

Ponti OMB De, Inggamer H (1976) Technical note: an improved leaf disk technique for biotests. Euphytica 25:129–130

Poos FW, Smith FF (1931) A comparison of oviposition and nymphal development, *Empoasca fabae* (Harris) on different host plants. J Econ Entomol 24:361–371

Pope WK, Dewey WG (1979) Breeding strategies for handling quantitative resistances to plant diseases. In: Proc Symp 5th Int Wheat Genet New Delhi 2:1074–1078

Porter KB, Daniels NE (1963) Inheritance and heritability of greenbug resistance in a common wheat cross. Crop Sci 3:116–118

Porter KB, Peterson GL, Vise O (1982) A new greenbug biotype. Crop Sci 22:847–850

Potrykus I (1984) Direkter Gentransfer von *Escherichia coli* zu *Nicotiana tabacum*, p 60. Univ Bonn

Power JB, Cummins SE, Cocking EC (1970) Fusion of isolated plant protoplasts. Nature (London) 225:1016–1018

Powers HR Jr, Sando WJ (1957) Genetics of host-parasite relationship in powdery mildew of wheat. Phytopathology 47:453 (Abstr)

Prakasa Rao PS, Rao YS, Israel P (1971) Factors favouring incidence of rice pests and methods of forecasting outbreaks: Gall midge and stem borers. Oryza 8:337–344

Prasada R, Sharma SK, Merchanela MM (1964) Identification of new biotype within Indian race 162 of *Puccinia recondita* Rob. ex. Desm. Ind Phytopathol 17:335–336

Prasada Rao U, Kalode MB, Srinivasan TE, Seshu DV (1977) Resistance of brown planthopper in rice. Ind J Genet Plant Breed 37:453–456

Pringle RB, Scheffer RP (1964) Host-specific plant toxins. Annu Rev Phytopathol 2:133–156

Provvidenti R, Gonsalves D, Ranalli P (1982) Inheritance of resistance to soybean mosaic virus in *Phaseolus vulgaris*. J Hered 73:302–303

Provvidenti R, Gonsalves D, Taiwo MA (1983) Inheritance of resistance to blackeye cowpea mosaic and cowpea aphid borne mosaic virus in *Phaseolus vulgaris*. J Hered 74:60–61

Puhalla JE (1969) The formation of diploids of *Ustilago maydis* on agar medium. Phytopathology 59:1771–1772

Purkayastha RP, Raha C (1983) Role of chlorogenic acid in the disease resistance of jute. Ind J Plant Pathol 1:49–55

Purushothaman D (1975) Changes in phenolic compounds in rice varieties as influenced by *Xanthomonas oryzae*. IL RISO XXV:85–89

Purushothaman D, Prasad NN (1972) Isolation of a toxin from *Xanthomonas oryzae*. Phytopathol Z 75:178–180

Putt ED (1964) Breeding behaviour of resistance to leaf mottle in sunflower. Crop Sci 4:177–179

Qualset CO (1975) Sampling germplasm in a center of diversity: an example of disease resistance in Ethiopian barley. In: Frankel OH, Hawkes JG (eds) Crop genetic resources for today and tomorrow. Cambridge Univ Press, Cambridge, pp 81–96

Radcliffe EB, Lauer FI (1970) Further studies on resistance to green peach aphid and potato aphid in wild tuber bearing *Solanum* species. J Econ Entomol 63:110–114

Rahe J, Kuc J, Chuang C, Williams E (1969) Induced resistance in *Phaseolus vulgaris* to bean anthracnose. Phytopathology 59:1641–1645

Raj S, Patel PN (1978) Studies on resistance in crops to bacterial diseases in India. X. Inheritance of multiple disease resistance in cowpea. Ind Phytopathol 31:294–299

Rajaratnam JH, Hock L (1975) Effect of boron nutrition on intensity of red spider mite attack on oil palm seedlings. Exp Agric 11:59–63

Ram HH, Singh RD, Singh YV (1981) Note on inheritance of resistance to powdery mildew and days to flowering in peas. Curr Sci 50:782–784

Ramey HH (1962) Genetics of plant pubescence in Upland cottons. Crop Sci 2:269

Rana BS, Tripathi DP, Rao NGP (1976) Genetic analysis of Exotic × Indian crosses in sorghum. XV. Inheritance of resistance to sorghum rust. Ind J Genet Plant Breed 36:244–249

Rana BS, Anahosur KH, Rao VJM, Parameshwarappa R, Rao NGP (1982) Inheritance of field resistance to sorghum charcoal rot and selection for multiple disease resistance. Ind J Genet Plant Breed 42:302–310

Rao AP, Fleming AA (1980) Cytoplasmic effects on northern leaf blight of maize. Ind J Genet Plant Breed 40:285–289

Rao MV (1968) Control of plant diseases-some possible approaches. Ind J Genet Plant Breed 28:128–141

Rao UP, Seetharaman R (1980) Screening of donors for brown planthopper in rice. Ind J Genet Plant Breed 40:241–246

Rasid G, Quick JS, Statler GD (1976) Inheritance of leaf rust resistance in three durum wheats. Crop Sci 16:294–296

Rastogi KB, Saini SS (1984) Inheritance of resistance to pea blight *(Ascochyta pinodella)* and induction of resistance in pea (*Pisum sativum* L.). Euphytica 33:9–11

Rautela GS, Payne MG (1970) The relationship of peroxidase and orthodiphenol oxidase to resistance of sugarbeets to *Cercospora* leaf spot. Phytopathology 60:238–245

Reddick D (1939) Scab immunity. Am Potato J 16:71–76

Reddy APK (1980) A report on bacterial leaf blight epidemics in Punjab state. All India Coordinated Rice Improvement Project, Mimeo

Reddy MSS, Rao MV (1979) Resistance genes and their deployment for control of leaf rust of wheat. Ind J Genet Plant Breed 39:359–365

Ree JH (1971) Induced mutations for rice improvement in Korea. In: Proc Symp Rice Breed Induced Mutat III. Vienna Int At Energy Ag Tech Rep Ser No 131:131–147

Reeder BD, Norton JD, Chambliss OL (1972) Inheritance of bean yellow mosaic virus resistance in southern pea, *Vigna sinensis* (Torner) S. J Am Soc Hortic Sci 97:235–237

Reeleder RD, Hagedorn DJ (1981) Inheritance of resistance to *Pythium myriotylum* hypocotyl rot in *Phaseolus vulgaris* L. Plant Dis 65:427–429

Reifschneider FJB, Arny DC (1983) Inheritance of resistance in maize to *Kabatiella zeae*. Crop Sci 23:615–616

Rick CM (1967) Exploiting species hybrids for vegetable improvement. Proc 17th Int Hortic Congr 3:217–229

Riley R, Chapman V, Johonsen R (1968) Introduction of yellow rust resistance of *Aegilops comosa* into wheat by genetically induced homoeologous recombination. Nature (London) 217:383–384

Ringlund K, Everson EH (1968) Leaf pubescence in common wheat *Triticum aestivum* L. and resistance to the cereal leaf beetle, *Oulema melanopus* (L.). Crop Sci 8:705–710

Roane CW (1973) Trends in breeding for disease resistance in crop. Annu Rev Phytopathol 11:463–486

Roane CW, Tolin SA, Genter CF (1983) Inheritance of resistance to maize dwarf mosaic virus in maize inbred line Oh7B. Phytopathology 73:845–850

Roberts CL, Staten G (1972) Heritability of *Verticillium* wilt tolerance in crosses of American upland cotton. Crop Sci 12:63–66

Roberts DD (1981) Sources of resistance to *Puccinia menthae* in Mint. Plant Dis 65:322–324

Roberts DWA (1954) Sawfly resistance in wheat. I. Types of resistance. Can J Agric Sci 34:582–597

Roberts JJ, Patterson FL, Gallun RL, Hendricks LT, Finey RE, Shaner GE, Ohm HW (1976) Downy soft red winter wheat, resistant to the cereal leaf beetle. Purdue Univ Agric Exp Stn Bull 115:3

Roberts JJ, Gallun RL, Patterson FL, Foster JE (1979) Effects of wheat leaf pubescence on hessian-fly. J Econ Entomol 72:211–214

Robertson DS, Walter EV (1963) Genetic studies of ear worm resistance in maize utilizing a series of chromosome nine translocations. J Hered 54:267–272

Robinson RA (1969) Disease resistance terminology. Rev Appl Mycol 48:593–606

Robinson RA (1971) Vertical resistance. Rev Plant Pathol 50:233–239

Robinson RA (1976) Plant pathosystems. Advanced series in agricultural sciences. Springer, Berlin Heidelberg New York, pp 184

Robinson SH, Wolfenbarger DA, Dilday RH (1980) Antixenosis of smooth leaf cotton to the ovipositional response of tobacco budworm. Crop Sci 20:646–649

Rockefeller Found Program Agric Sci (1963) Annu Rep 1962–63:310

Rockefeller Found Program Agric Sci (1964) Annu Rep 1963–64:285

Rodriguez M, Galvez GE (1975) Indications of partial resistance of rice to the fungus *Pyricularia oryzae* Cav. In: Galvez GE (ed) Proc Semin Horizontal Resist Blast Dis Rice. Ser CE-9. CIAT, Cali, Colombia, pp 137–154

Rogers CE (1981) Resistance of sunflower species to the western potato leafhopper. Environ Entomol 10:697–700

Rogers CE, Thompson TE (1978a) Resistance in wild *Helianthus* to the sunflower beetle. J Econ Entomol 71:622–623

Rogers CE, Thompson TE (1978b) *Helianthus* resistance to the carrot beetle. J Econ Entomol 71:760–761

Rogers KM, Norton JD, Chambliss OL (1973) Inheritance of resistance to cowpea chlorotic mottle virus in southern pea, *Vigna sinensis*. J Am Soc Hortic Sci 98:62–63

Rohringer R, Howes NK, Kim WK, Samborski DJ (1974) Evidence for a gene-specific RNA determining resistance in wheat to stem rust. Nature (London) 249:585–588

Rojanaridpiched C, Gracen VE, Everett HL, Coors JG, Pugh BF, Bouthyette P (1984) Multiple factor resistance in maize to European corn borer. Maydica XXIX:305–315

Romero S, Erwin DC (1969) Variation in pathogenicity among single oospore cultures of *Phytophthora infestans*. Phytopathology 59:1310–1317

Romig RW, Caldwell RM (1968) Stomatal exclusion of *Puccinia recondita* by wheat peducles and sheaths. Phytopathology 54:214–218

Rosielle AA, Brown AGP (1979) Inheritance heritability and breeding behaviour of three sources of resistance to *Septoria tritici* in wheat. Euphytica 28:392

Ross H (1966) The use of wild *Solanum* species in German potato breeding of past and today. Am Potato J 43:63

Rouse DI. Nelson RR, MacKenzie DR, Armitage CR (1980) Analysis of some components of horizontal resistance and parasitic fitness in four wheat cultivars and six isolates of *Erysiphe graminis* f.sp. *tritici*. Phytopathology 70:1097–1100

Rowe DF, Hill RR Jr (1981) Inter-population improvement procedure for alfalfa. Crop Sci 21: 392–397

Rudorf W (1950) Methods and results of breeding resistant strains of potatoes. Am Potato J 27: 332–339

Russell GE (1964) Breeding for tolerance to beet yellows virus and beet mild yellowing virus in sugarbeet. I. Selection of breeding methods. Ann Appl Biol 53:363

Russell GE (1966) Observations on the seedling behaviour of aphids on sugarbeet plants in the glasshouse. J Agric Sci 67:405–410

Russell GE (1978) Plant breeding for pest and disease resistance. Butterworth, London, p 485

Sackston WE (1956) Observation and speculations on rust (*Puccinia helianthi* Schw.) and some other diseases of sunflower in Chile. Plant Dis Rep 40:744–747

Sackston WE (1957) Diseases of sunflower in Uruguay. Plant Dis Rep 41:885–889

Sackston WE (1962) Studies on sunflower rust. III. Occurrence, distribution and significance of races of *Puccinia helianthi* Schw. Can J Bot 40:1449–1458

Safeeulla KM (1977) Genetic vulnerability: The basis of recent epidemics in India. Ann NY Acad Sci 287:72–85

Sakai R, Toriyama K (1964) Relation between factors related to phenols and varietal resistance of potato plant to late blight. I. Field observation. Ann Phytopathol Soc Jpn 29:33–38

Saleman RN (1931) Recents progress dans la creation de varieties de pommes de terre resistant au mildiou "*Phytophthora infestans*". 10th Congr Int Pathol Comp 2:435

Samaddar KR, Scheffer RP (1970) Effects of *Helminthosporium victoriae* toxin on germination and aleurone secretion by resistant and susceptible seeds. Plant Physiol 45:586–590

Samborski DJ (1963) A mutation in *Puccinia recondita* Rob. ex. Desm. f. sp. *tritici* to virulence on Transfer, Chinese spring × *Aegilops umbellulata* Zhuk. Can J Bot 41:475–479

Samborski DJ, Dyck PL (1968) Inheritance of virulence in wheat leaf rust on the standard differential wheat varieties. Can J Genet Cytol 10:24–32

Sams DW, Lauer FI, Radcliffe EB (1975) Excised leaflet test for evaluating resistance to green peach aphid in tuber-bearing *Solanum* germplasm. J Econ Entomol 68:607–609

Sams DW, Lauer FI, Radcliffe EB (1976) Breeding behaviour of resistance to green peach aphid in tuber-bearing *Solanum* germplasm. Am Potato J 53:23–29

Sanford LL, Ladd TL (1983) Selection for resistance to potato leafhopper in potatoes. III. Comparison of two selection procedures. Am Potato J 69:653–659

Sanghi AK (1974) Genetic relationship of resistance possessed by four wheat cultivars to stem rust strains with unusual genes for avirulence. Ind J Genet Plant Breed 34:384–389

Sanz MP (1981) Bacterial induction of the accumulation of phaseollin, pisatin and rishitin and their antibacterial activity. Neth J Plant Pathol 87:119–129

Sasamoto K (1961) Resistance of the rice plant with silicate and nitrogenous fertilizers to the rice stem borer, *Chilo suppressalis* (Walker). Proc Fac Liberal Arts Educ Yamanasaki Univ Jpn No 3: 1–73

Satyanarayanaiah K, Reddy MV (1972) Inheritance of resistance to insect gall-midge (*Pachydiplosis oryzae* Wood Mason) in rice. Andhra Agric J 19:1–8

Savitsky H, Price C (1965) Resistance to the sugarbeet nematode *(Heterodera schachti)* in F_1 hybrids between *Beta vulgaris* and *Beta patellaris*. J Am Soc Sugarbeet Technol 13:370–373

Saxena JK, Tripathi RM, Srivastava RL (1975) Powdery mildew resistance in pea *(Pisum sativum* L.). Curr Sci 44:746

Saxena KMS, Hooker AL (1968) On the structure of a gene for disease resistance in maize. Proc Natl Acad Sci 61:1300–1305

Saxena KN (1969) Patterns of insect-plant relationships determining susceptibility or resistance of different plants to an insect. Entomol Exp Appl 12:751–766

Saxena KN, Gandhi JR, Saxena RC (1974) Patterns of relationships between certain leafhoppers and plants. Entomol Exp Appl 17:303–318

Saxena RC, Pathak MD (1977) Factors affecting resistance of rice varieties to the brown planthopper, *Nilaparvata lugens* (Stal.). In: Pest Control Counc Philippines, Bacolod, p 34

Schafer JE, Caldwell RM, Patterson FL, Campton LE (1963) Wheat leaf rust resistance combinations. Phytopathology 53:569–573

Schafer JF (1971) Tolerance to plant disease. Annu Rev Phytopathol 9:235–252

Scheffer RP, Nelson RR, Ullstrup AJ (1967) Inheritance of toxin production and pathogenicity in *Cochliobolus carbonum* and *Cochliobolus victoriae*. Phytopathology 57:1288–1291

Scheifele GL, Nelson RR, Koons C (1969) Male sterility cytoplasm conditioning susceptibility of resistant inbred lines of maize to yellow leaf blight caused by *Phyllosticta zeae*. Plant Dis Rep 53:656–659

Schertz KF, Tai YP (1969) Inheritance of reaction of *Sorghum bicolor* (L.) Moench to toxin produced by *Periconia circinata* (Mang.) Sacc. Crop Sci 9:621–624

Schick R, Schick E, Haussdorfer M (1958a) Ein Beitrag zur physiologischen Spezialisierung von *Phytophthora infestans*. Phytopathol Z 31:225–236

Schick R, Moller KH, Haussdorfer M, Schick E (1958b) Die Widerstandsfähigkeit von Kartoffelsorten gegenüber der durch *Phytophthora infestans* (Mont.) de Bary hervorgerufenen Krautfäule. Zuechter 28:99–105

Schillinger JA (1976) Host plant resistance to insects in soybeans. In: Proc World Soybean Conf I, Illinois, pp 579–583

Schillinger JA, Gallun RA (1968) Leaf pubescence of wheat as a deterrent to the cereal leaf beetle, *Oulema melanopus*. Ann Entomol Soc Am 61:900–903

Schlosberg J, Baker WA (1948) Tests of sweet corn lines for resistance to European corn borer larvae. J Agric Res 77:137–159

Schnathorst WC, Vay JE de (1963) Common antigens in *Xanthomonas malvacearum* and *Gossypium hirsutum* and their possible relationship to host specificity and disease resistance. Phytopathology 53:1142

Schoonhaven AV (1974) Resistance to thrips damage in cassava. J Econ Entomol 67:728–730

Schuster ML, Coyne DP (1981) Biology, epidemiology, genetics and breeding for resistance to bacterial pathogens of *Phaseolus vulgaris* L. Hortl Rev 3:1–58

Schuster MF, Maxwell FG, Jenkins JN, Cherry ET, Parrott WL, Holder DG (1973) Resistance to two-spotted spider mite in cotton. Miss Agric Exp Stn Bull 802

Schwartz HF, Pastor Corrales MA, Singh SP (1982) New sources of resistance to anthracnose and angular leaf spot of beans *(Phaseolus vulgaris* L.). Euphytica 31:741–754

Schwinghamer EA (1959) The relation between radiation dose and the frequency of mutations for pathogenicity in *Melampsora lini*. Phytopathology 49:260–269

Scott GE, Guthrie WD (1967) Reaction of permutations of maize double crosses to leaf feeding of European corn borers. Crop Sci 7:233–235

Scott GE, Hallauer AR, Dicke FF (1964) Types of gene actions conditioning resistance to European corn borer leaf feeding. Crop Sci 4:603–606

Scott GE, Dicke FF, Penny LH (1965) Effects of first brood European corn borers on single crosses grown at different nitrogen and plant population levels. Crop Sci 5:261–263

Searls EM (1935) The relation of foliage colour to aphid resistance in some varieties of canning peas. J Agric Res 51:613–619

Sears ER (1956) The transfer of leaf rust resistance from *Aegilops umbellulata* to wheat. Brookhaven Symp Biol 9:1–22

Sears ER (1966) Chromosome mapping with the aid of telocentrics. Proc 2nd Int Wheat Genet Symp, Hereditas, Suppl 2:370–381

Sebastian SA, Risius ML, Royer MH (1983) Inheritance of powdery mildew resistance in wheat line IL 72–2219–1. Plant Dis 67:943–945

Seetharaman R, Prasad K, Anajaneyulu (1976) Inheritance of resistance to rice tungro virus disease. Ind J Genet Plant Breed 36:34–37

Seetharaman R, Krishnaiah NV, Sobha Rani N, Kalode MB (1984) Studies on varietal resistance to brown planthopper in rice. Ind J Genet Plant Breed 44:65–72

Seevers PM, Daly JM (1970) Studies on wheat stem rust resistance controlled at the Sr6 locus. II. Peroxidase activities. Phytopathology 60:1642–1647

Selim AK, Serry MS, Omran AO, Satour MM (1976a) Breeding for disease resistance in sesame (Sesamum indicum L.). II. Inheritance of resistance to root and stem rot disease caused by Sclerotium bataticola Taub. Egypt J Phytopathol 8:15–18

Selim AK, Serry MS, Satour MM, Alahmar BA (1976b) Breeding for resistance in sesame (Sesamum indicum L.). III. Inheritance of resistance to Fusarium oxysporum Schlecht. Egypt J Phytopathol 8:19–24

Sen Gupta GC (1969) The recognition of biotypes of the woolly aphid, Eriosoma lanigerum (Hausmann) in sourth Australia by their differential ability to colonise varieties of apple rootstock, and an investigation of some possible factors in the susceptibility of varieties to those insects. PhD thesis, Waite Agric Res Inst, Adelaide

Sen Gupta GC, Miles PW (1975) Studies on the susceptibility of varieties of apple to the feeding of two strains of woolly aphids (Homoptera) in relation to the chemical content of the tissues of the host. Aust J Agric Res 26:157–168

Serry MS, Selim AK, Satour MM, Alahmar BA (1976) Breeding for disease resistance in sesame (Sesamum indicum L.). I. Inheritance of resistance to Rhizoctonia root rot. Egypt J Phytopathol 8:9–14

Shade RE, Thompson TE, Campbell WR (1975) An alfalfa weevil larval resistance mechanism detected in Medicago. J Econ Entomol 68:399–404

Shands RG, Cartwright WB (1953) A fifth gene conditioning hessian-fly response in common wheat. Agron J 45:302–307

Shaner G (1973a) Estimation of conidia production by individual pustules of Erysiphe graminis f.sp. tritici. Phytopathology 63:847–850

Shaner G (1973b) Evaluation of slow-mildewing resistance of Knox wheat in the field. Phytopathology 63:867–872

Shaner G (1973c) Reduced infectibility and inoculum production as factors of slow mildewing in Knox wheat. Phytopathology 63:1307–1311

Shaner G, Hess FD (1978) Equations for integrating components of slow leaf-rusting resistance in wheat. Phytopathology 68:1464–1469

Shaner G, Ohm HW, Finney RE (1978) Response of susceptible and slow leaf-rusting wheats to infection by Puccinia recondita. Phytopathology 68:471–475

Sharma BR, Swarup V, Chatterjee SS (1972) Inheritance of resistance to black rot in cauliflower. Can J Genet Cytol 14:363–370

Sharma D, Knott DR (1966) The transfer of leaf-rust resistance from Agropyron to Triticum by irradiation. Can Genet Cytol 8:137–143

Sharma D, Gupta SC, Rai GS, Reddy MV (1984) Inheritance of resistance to sterility mosaic disease in pigeonpea. I. Ind J Genet Plant Breed 44:84–90

Sharma GS, Singh RB (1975) Reaction of wheat varieties to rusts. Ind J Genet Plant Breed 35:146–151

Sharma HC, Gill BS, Uyemote JK (1984) High levels of resistance in Agropyron species to barley yellow dwarf and wheat streak mosaic virus. Phytopath Z 110:143–147

Sharma SK, Mishra DP (1965) Mutation in race 107 of brown rust of wheat, Puccinia recondita. Ind Phytopathol 18:363–366

Sharma SK, Prasad R (1970) Somatic recombination in the leaf rust of wheat caused by Puccinia recondita Rob. ex. Desm. Phytopathol Z 67:240–244

Sharma VK, Chatterji SM (1971) Preferential oviposition and antibiosis of different maize germplasms against Chilo zonellus (Sum.) under cage conditions. Ind J Entomol 33:299–311

Shastry MVS, Prakasa Rao PS (1973) Inheritance of resistance to rice gall midge, *Pachydiplosis oryzae* Wood. Mason. Curr Sci 42:652–653

Shastry SVS, Sharma SD, John VT, Krishnaiah K (1971) New sources of resistance to pests and diseases in the Assam rice collections. Int Rice Comm Newsl XX(3):1–16

Shastry SVS, Freeman WH, Seshu DV, Israel P, Roy JK (1972) Host-plant resistance to rice gall midge. In: Rice breeding. Int Rice Res Inst, Philippines, pp 353–365

Shaver TN, Lukefahr MJ (1971) A bioassay technique for detecting resistance of cotton strains to tobacco bud worms. J Econ Entomol 64:1274–1277

Shaver TN, Lukefahr MJ, Garcia JA (1970) Food utilization, ingestion, and growth of larvae of the bollworm and tobacco budworm on diets containing gossypol. J Econ Entomol 63:1544–1546

Shepard JE (1981) Protoplasts as sources of disease resistance in plants. Annu Rev Phytopathol 19:145–166

Shepherd KW, Mayo GME (1972) Genes conferring specific plant disease resistance. Science 175:375–380

Shukla GP, Pandya BP, Singh DP (1978) Inheritance of resistance to yellow mosaic in mungbean. Ind J Genet Plant Breed 38:357–360

Sidhu G, Person C (1971) Genetic control of virulence in *Ustilago hordei* II. Segregations for higher levels of virulence. Can J Genet Cytol 13:173–178

Sidhu G, Person C (1972) Genetic control of virulence in *Ustilago hordei*. III. Identification of genes for host resistance and demonstration of gene-for-gene relations. Can J Genet Cytol 14:209–213

Sidhu GS, Khush GS (1978) Dominance reversal of bacterial blight resistance in some rice varieties. Phytopathology 68:461–463

Sidhu GS, Khush GS (1979) Linkage relationships of some genes for disease and insect resistance and semidwarf stature in rice. Euphytica 28:233–237

Sidhu GS, Khush GS, Medrano FG (1979) A dominant gene in rice for resistance to white-backed planthopper and its relationship with other characters. Euphytica 28:227–232

Sifuentes JA, Painter RH (1964) Inheritance of resistance to western corn root worm adults in field corn. J Econ Entomol 57:475–477

Sigurbjornsson B, Micke A (1969) Progress in mutation breeding. In: Proc Symp Induced Mutat Plants. Int At Energy Ag, Vienna, pp 673–698

Sikora RA, Koshy PK, Malek RB (1972) Evaluation of wheat selections for resistance to cereal leaf beetle. Ind J Nematol 2:81–82

Simmonds NW (1962) Variability in crop plants, its use and conservation. Biol Rev Cambridge Philos Soc 37:422–465

Simons MD (1965) Relationship between the response of oats to crown rust and kernel density. Phytopathology 55:579–582

Simons MD (1966) Relative tolerance of oat varieties to the crown rust fungus. Phytopathology 56:36–40

Simons MD (1968) Additional sources of tolerance to oat crown rust. Plant Dis Rep 52:59–61

Simons MD (1969) Heritability of crown rust tolerance in oats. Phytopathology 59:1329–1333

Simons MD (1971) Modification of tolerance of oats to crown rust by mutation induced with ethyl methane sulfonate. Phytopathology 61:1064–1067

Simons MD (1972a) Polygenic resistance to plant disease and its use in breeding resistant cultivars. J Environ Qual 1:232–240

Simons MD (1972b) Crown rust tolerance in *Avena sativa* type oats derived from wild *Avena sterilis*. Phytopathology 63:1444–1446

Simons MD (1975) Heritability of field resistance to the oat crown rust fungus. Phytopathology 65:324–328

Simons MD (1985) Transfer of field resistance to *Puccinia coronata* from *Avena sterilis* to cultivated oats by backcrossing. Phytopathology 75:314–317

Simons MD, Browning JA (1961) Seed weight as a measure of response of oats to crown rust infection. Proc Iowa Acad Sci 68:114–118

Simons MD, Murphy HC (1967) Determination of relative tolerance of *Puccinia coronata* avenae of experimental lines of oats. Plant Dis Rep 41:947–950

Sindhu JS, Singh KP, Slinkard AE (1983) Inheritance of resistance to *Fusarium* wilt in chickpeas. J Hered 74:68

Singh AK, Saini SS (1980) Inheritance of resistance to angular leaf spot (*Isariopsis griseola* Sacc.) in frenchbean (*Phaseolus vulgaris* L.). Euphytica 29:175–176

Singh BB, Malick AS (1978) Inheritance of resistance to yellow mosaic in soybean. Ind J Genet Plant Breed 38:258–261

Singh BB, Gupta SC, Singh BD (1974a) Sources of field resistance to rust and yellow mosaic diseases of soybean. Ind J Genet Plant Breed 34:400–404

Singh BB, Singh BD, Gupta SC (1974b) PI 191443 and *Glycine formosana* – resistant sources of yellow mosaic. Soybean Genet Newsl 1:17–18

Singh CB, Rao YP (1971) Association between resistance to *Xanthomonas oryzae* and morphological and quality characters induced mutants of *indica* and *japonica* varieties of rice. Ind J Genet Plant Breed 31:369–373

Singh D, Patel PN (1977) Studies on resistance in crops to bacterial diseases in India. VII. Inheritance of resistance to bacterial blight disease in cowpea. Ind Phytopathol 30:99–102

Singh DP (1980) Inheritance of resistance to yellow mosaic virus in blackgram (*Vigna mungo* (L.) Hepper). Theor Appl Genet 57:233–235

Singh DP (1981) Breeding for resistance to diseases in greengram and blackgram. Theor Appl Genet 59:1–10

Singh DP (1982) Breeding for resistance to mungbean yellow mosaic virus in mungbean and urdbean. Ind J Genet Plant Breed 42:348–355

Singh DP, Nanda JS (1975) Inheritance of resistance to bacterial leaf blight in rice. J Hered 66:384–386

Singh DP, Nanda JS (1980) Genetics of rice tungro virus and its vector. IL RISO XXIX(1):43–51

Singh DP, Sharma BL (1983) Inheritance of resistance to yellow mosaic virus in *Vigna radiata* X *V. radiata* var. *sublobata*. 15th Int Congr Genet Abstr No 1331:735

Singh IS, Asnani VL (1975) Gene effects for resistance to brown stripe downy mildew in maize. Ind J Genet Plant Breed 35:123–127

Singh KB, Reddy MV (1983) Inheritance of resistance to *Ascochyta* blight in chickpea. Crop Sci 23:9–10

Singh KB, Hawtin GC, Nene YL, Reddy MV (1981) Resistance of chickpeas to *Ascochyta rabiei*. Plant Dis 65:586–587

Singh R (1953) Inheritance in maize of reaction to the European corn borer. Ind J Genet Plant Breed 13:18–47

Singh RJ, Khush GS, Mew TW (1983) A new gene for resistance to bacterial blight in rice. Crop Sci 23:558–560

Singh TB, Singh RA, Singh BN (1983a) Genetics of pathogenicity of *Helminthosporium maydis*. 15th Int Congr Genet Abstr No 1333:736

Singh TB, Singh RA, Singh KMP, Faruqui OR (1983b) Genetics of pathogenicity in *Helminthosporium sativum*. 15th Int Congr Genet Abstr No 1334:736

Singh VS, Bhatia SK (1979) Inheritance of resistance to rice weevil. Proc 15th Int Wheat Genet Symp 2:1105–1113

Sitterly WR (1972) Breeding for disease resistance in cucurbits. Annu Rev Phytopathol 10:471–490

Sittiyos P, Poehlman JM, Sehgal OP (1979) Inheritance of resistance to cucumber mosaic virus in mungbean. Crop Sci 19:51–53

Siwi BH, Khush GS (1977) New genes for resistance to green leafhopper in rice. Crop Sci 17:17–20

Skinner DZ, Stuteville DL (1985) Genetics of host-parasite interactions between alfalfa and *Peronospora trifoliorum*. Phytopathology 75:119–121

Skrdla WH, Alexander LJ, Oakes G, Dodge AF (1968) Horticultural characters and reaction of two diseases of the world collection of the genus *Lycopersicon*. Ohio Agric Exp Stn Res Bull 1009:110

Slesinski RS, Ellingboe AH (1970) Gene-for-gene interactions during primary infection of wheat by *Erysiphe graminis* f.sp. *tritici*. Phytopathology 60:1068–1070

Slykhuis JT (1955) *Aceria tulipae* Keifer (Acarina:Eriophyide) in relation to spread of wheat streak mosaic. Phytopathology 45:116–128

Smith CM, Brim CA (1979) Field and laboratory evaluation of soybean lines for resistance to corn earworm leaf feeding. J Econ Entomol 72:78–80

Smith DG, McInnes AG, Higgins VJ, Miller RL (1971) Nature of the phytoalexin produced by alfalfa in response to fungal infection. Physiol Plant Pathol 1:41–44

Smith DH, Webster JA (1973) Resistance to cereal leaf beetle in Hope substitution lines. In: Proc Symp 4th Int Wheat Genet, Columbia, Mo, pp 761–764

Smith DH, Webster JA, Everson EH (1978) Registration of nine germplasm lines of hard red winter wheat. Crop Sci 18:166

Smith HH (1974) Model systems for somatic cell plant genetics. Bioscience 24:269–276

Smith OD, Schlehuber AM, Curtis BC (1962) Inheritance studies of greenbug resistance in four varieties of winter barley. Crop Sci 2:489–491

Snelling RO (1941a) The place and methods of breeding for insect resistance in cultivated plants. J Econ Entomol 34:335–340

Snelling RO (1941b) Resistance of plants to insect attack. Bot Rev 7:543–586

Sogawa K, Pathak MD (1970) Mechanisms of brown planthopper resistance in Mudgo variety of rice (Hemiptera:Delphacidae). Appl Entomol Zool 5:145–158

Sokhi SS, Joshi LM, Rao MV (1973) Inheritance of *Alternaria* resistance in wheat. Ind J Genet Plant Breed 33:457–459

Sosa O Jr (1978) Biotype L, ninth biotype of the hessian-fly. J Econ Entomol 71:458–460

Sosa O Jr (1981) Biotypes J and L of the hessian-fly discovered in an Indiana wheat field. J Econ Entomol 74:180–181

Soto PE (1974) Ovipositional preference and antibiosis in relation to resistance to a sorghum shoot-fly. J Econ Entomol 67:256–257

Sowell G Jr (1981) Additional sources of resistance to gummy stem blight of muskmelon. Plant Dis 65:253–254

Sprague GF (1980) Germplasm resources of plants. Their preservation and use. Annu Rev Phytopathol 18:147–165

Srinivasachar D, Verma PK (1971) Induced aphid resistance in *Brassica juncea* Coss. Curr Sci 40:311–313

Srivastava DN (1972) Bacterial blight of rice. Ind Phytopathol 25:1–16

Srivastava OP (1982) Genetics of seedling resistance to leaf blight in wheat. Ind J Genet Plant Breed 42:140–141

Srivastava SN, Srivastava DP (1977) Inheritance of resistance to covered smut of barley. Ind J Genet Plant Breed 37:321–327

Stakman EC (1914) A study in cereal rusts. Physiologic races. Minn Agric Exp Stn Bull 138

Stakman EC (1968) The need for international testing centres for crop pests and pathogens. Bull Entomol Soc Am 14:124–128

Stakman EC, Christensen JJ (1960) The problem of breeding resistant varieties. In: Horsfall JE, Dimond AE (eds) Plant pathology, vol III. Academic Press, London New York, pp 567–624

Stakman EC, Harrar JG (1957) Principles of plant pathology. Ronald Press, New York

Stanghellini ME, Aragaki M (1966) Relation of periderm formation and callose deposition to anthracnose resistance in papaya fruit. Phytopathology 56:444–450

Starks KJ, Burton RL (1977) Greenbugs: A comparison of mobility on resistant and susceptible varieties of four small grains. Environ Entomol 6:331–332

Starks KJ, Merkle OG (1977) Low level resistance in wheat to greenbug. J Econ Entomol 70:305–306

Starks KJ, Eberhart SA, Doggett H (1970) Recovery from short fly attack in a sorghum diallel. Crop Sci 10:519–522

Starks KJ, Muniappan R, Eikenbary RD (1972a) Interaction between plant resistance and parasitism against the greenbug on barley and sorghum. Ann Entomol Soc Am 65:650–655

Starks KJ, Weibel DE, Johnson JW (1972b) Sorghum resistance to greenbug. Sorghum Newsl 15:10

Starks KJ, Casady AJ, Merkle OG, Boozaya Angoon D (1982) Chinch bug resistance in pearl millet. J Econ Entomol 75:337–339

Statler GD (1973) Inheritance of resistance to leaf rust in Waldron wheat. Phytopathology 63:346–348

Statler GD, Watkins JE, Nordgaard J (1977) General resistance displayed by three hard red spring wheat *(Triticum aestivum)* cultivars to leaf rust. Phytopathology 67:759–762

Statler GD, Hammond JJ, Zimmer DE (1981) Hybridization of *Melampsora lini* to identify rust resistance in flax. Crop Sci 21:219–221

Stavely JR, Pittarelli GW, Burk LG (1973) *Nicotiana repanda* as a potential source for disease resistance in *N. tabacum*. J Hered 64:265–271

Stebbins NB, Patterson FL, Gallun RL (1980) Interrelationships among wheat genes for resistance to hessian-fly. Crop Sci 20:177–180

Stebbins NB, Patterson FL, Gallun RL (1982) Interrelationships among wheat genes H_3, H_6, H_9, and H_{10} for hessian-fly resistance. Crop Sci 22:1029–1032

Stebbins NB, Patterson FL, Gallun RL (1983) Inheritance of resistance to PI 94587 wheat to biotypes B and D of hessian-fly. Crop Sci 23:251–253

Steffensen BJ, Wilcoxson RD, Roelfs AP (1984) Inheritance of resistance to *Puccinia graminis* f.sp. *secalis*. Plant Dis 68:762–763

Steiner GW, Byther RS (1971) Partial characterization and use of a host specific toxin from *Helminthosporium sacchari* on sugarcane. Phytopathology 61:691–695

Stephens SG (1959) Laboratory studies on feeding and oviposition preferences of *Anthonomus grandis* Boh. J Econ Entomol 52:390–396

Stephens SG, Lee HS (1961) Further studies on feeding and oviposition preferences of the boll weevil (*Anthonomus grandis* Boh.). J Econ Entomol 54:1085–1090

Stevens RB (1949) Replanting "discarded" varieties as a means of disease control. Science 110:49

Stockwell V, Hanchy P (1984) The role of cuticle in resistance of beans to *Rhizoctonia solani*. Phytopathology 74:1640–1642

Stoessl A (1972) Inermin associated with pisatin in peas inoculated with the fungus *Monilinia fructicola*. Can J Biochem 50:107–108

Stoner AK (1970) Breeding for insect resistance in vegetables. Hortic Sci 5:76–79

Stoner AK, Frank JA, Gentile AG (1968) The relationship of glandular hairs on the tomatoes to spider mite resistance. Proc Am Soc Hortic Sci 93:532–538

Strobel G (1974) Phytotoxins produced by plant parasites. Annu Rev Plant Physiol 25:541–566

Sulladmath VV, Shivshankar G, Anil Kumar TB, Veerappa KB (1977) Inheritance of resistance to bacterial leaf spot in *Dolichos* crosses. Ind J Genet Plant Breed 37:101–102

Sunderland N, Xu ZH (1982) Shed pollen culture in *Hordeum vulgare*. J Exp Bot 33:1086–1095

Sunderwirth SD, Roelfs AP (1980) Greenhouse evaluation of the adult plant resistance of Sr2 to wheat stem rust. Phytopathology 70:634–637

Suneson CA (1960) Genetic diversity – a protection against plant diseases and insects. Agron J 52:319–321

Suneson CA, Noble WB (1950) Further differentiation of genetic factors in wheat for resistance to the hessianfly. US Dep Agric Tech Bull 1004

Swaminathan MS (1980) Past, present and future trends in tropical agriculture. In: Perspectives in World agriculture. CAB Farnham Royal (Slough), pp 1–48

Swarup V, Raghava SPS (1974) Induced mutation for resistance to leaf-curl virus and its inheritance in garden *Zinnia*. Ind J Genet Plant Breed 34:17–21

Taiwo MA, Provvindenti R, Gonsalves D (1981) Inheritance of resistance to blackeye cowpea mosaic virus in *Vigna unguiculata*. J Hered 72:433–434

Talekar NS, Chen BS (1983) Identification of sources of resistance to limabean pod borer (Lepidoptera:Pyralidae) in soybean. J Econ Entomol 76:38–39

Taleker NS, Lin YH (1981) Two sources with differing modes of resistance to *Callosobruchus chinensis* in mungbean. J Econ Entomol 74:639–642

Tatum LA (1971) The southern corn leaf blight epidemic. Science 171:1113–1116

Terry-Lewandowski VM, Stimart DP (1983) Multiple resistance in induced amphiploids of *Zinnia elegans* and *Z. angustifolia* to three major pathogens. Plant Dis 67:1387–1389

Thakur RP, Patel PN, Verma JP (1977) Genetical relationships between reactions to bacterial leaf spot, yellow mosaic and *Cercospora* leaf spot diseases in mungbean *(Vigna radiata)*. Euphytica 26:765–774

Thakur RP, Subba Rao KV, Williams RJ (1983a) Evaluation of a new field screening technique for smut resistance in pearl millet. Phytopathology 73:1255–1258

Thakur RP, Talukdar BS, Rao VP (1983b) Genetics of ergot resistance in pearl millet. 15th Int Congr Genet Abstr No 1336:737

Thanutong P, Furusawa I, Yamamoto M (1984) Resistant tobacco plants from protoplast-derived calluses selected for their resistance to *Pseudomonas* and *Alternaria* toxins. Theor Appl Genet 66:209–215

Thomas CA (1976) Resistance of VFR-1 safflower to *Phytophthora* root rot and its inheritance. Plant Dis Rep 60:123–125

Thomas CA (1981) Inheritance of resistance to lettuce mosaic virus in safflower. Phytopathology 71:817–818

Thomas JG, Sorensen EL, Painter RH (1966) Attached vs excised trifoliates for evaluation of resistance in alfalfa to the spotted alfalfa aphid. J Econ Entomol 59:444–448

Thomas PL (1982) Inheritance of virulence of *Ustilago nuda* on barley cultivars Warrior, Compana, Valkie, Keystone, and Bonanza. Can J Genet Cytol 25:53–56

Thomas PL, Metcalfe DR (1984) Loose smut resistance in two introductions of barley from Ethiopia. Can J Plant Sci 64:255–260

Thompson DL, Bergquist RR (1984) Inheritance of mature plant resistance to *Helminthosporium maydis* race 0 in maize. Crop Sci 24:807–811

Thompson KF (1963) Resistance to the cabbage aphid *(Brevicoryne brassicae)* in brassica plants. Nature (London) 198:209

Thompson TE, Shade RE, Axtell JD (1978) Alfalfa weevil resistance mechanism characterized by larval convulsions. Crop Sci 18:208–209

Thornton RJ, Johnsteon JR (1971) Rates of spontaneous mitotic recombination in *Saccharomyces cerevisiae*. Genet Res 18:147–151

Thung TH (1947) Potato diseases and hybridization. Phytopathology 37:373–381

Thyr BD (1968) Resistance to bacterial canker in tomato and its evaluation. Phytopathology 58: 279–281

Tingey WM, Laubengayer JE (1981) Defense against the green peach aphid and potato leafhopper by glandular trichomes of *Solanum berthaultii*. J Econ Entomol 74:721–725

Tinline RD (1962) *Cochliobolus sativus*. V. Heterokaryosis and parasexuality. Can J Bot 40:425–437

Tinline RD, Mac Neill BH (1969) Parasexuality in plant pathogenic fungi. Annu Rev Phytopathol 7:147–170

Tomiyama K, Sakuma T, Ishizaka N, Sato N, Katsui N, Takasugi M, Masamunc T (1968) A new antifungal substance isolated from resistant potato tuber tissue infected by pathogens. Phytopathology 58:115–116

Toriyama K (1975) Recent progress of studies on horizontal resistance in rice breeding for blast resistance in Japan. In: Proc Semin Horizontal Resist Blast Dis Rice. Centro Int Agric Trop, Cali Colombia

Torres E, Browning JA (1968) The yield of urediospores per unit of sporulating area as a possible measure of tolerance of oats to crown rust. Phytopathology 58:1070

Toxopeus HJ (1956) Reflections on the origin of new physiologic races of *Phytophthora infestans* and the breeding of resistance in potatoes. Euphytica 5:221–237

Toxopeus HJ (1957) On the influence of extra R-genes on the resistance of the potato to the corresponding P-races of *Phytophthora infestans*. Euphytica 6:106–110

Toynbee-Clarke G, Bond DA (1970) A laboratory technique for testing red clover seedlings for resistance to stem eelworm *(Ditylenchus dipsaci)*. Plant Pathol 19:173–176

Trenbath BR (1977) Interactions among diverse hosts and diverse pathogens. Ann NY Acad Sci 287:124–150

Trione EJ (1960) The HCN content of flax in relation to flax wilt resistance. Phytopathology 50: 482–486

Tripathi NN, Kaushik CD, Yadav TP, Yadav AK (1978) Studies on the inheritance of *Alternaria* blight resistance in raya. Ind Phytopathol 31:127

Tsai KH, Lu YC, Oka HI (1974) Mutation breeding of soybean for the resistance to rust disease. SABRAO J 6:181–191

Tseng CT, Guthrie WD, Russell WA, Robbins JC, Coats JR, Tollefson JJ (1984) Evaluation of two procedures to select for the European corn borer in a synthetic cultivar of maize. Crop Sci 24: 1129–1133

Tu JC, Beversdorf WD (1982) Tolerance to white mold (*Sclerotinia sclerotiorum* (Lib.) De Bary) in Ex Rico 23, a cultivar of white bean (*Phaseolus vulgaris* L.). Can J Plant Sci 62:65−69

Upadhyay MK, Kumar R (1979) Identification of genes governing seedling and adult plant resistance to Indian races of stripe rust of wheat. In: Proc Symp 5th Int Wheat Genet New Delhi 2: 1049−1056

Upadhyay MK, Prakash S (1980) Sources of resistance to powdery mildew of barley. Ind J Genet Plant Breed 40:18−21

Upadhyaya HD, Haware MP, Kumar J, Smithson JB (1983a) Resistance to wilt in chickpea. I. Inheritance of late wilting in response to race 1. Euphytica 32:447−452

Upadhyaya HD, Smithson JB, Haware MP, Kumar J (1983b) Resistance to wilt in chickpea. II. Further evidence for two genes for resistance to race 1. Euphytica 32:749−755

Vaidya GR, Kalode MB (1981) Studies on biology and varietal resistance of white backed planthopper, *Sogatella furcifera* (Horvath) in rice. Ind J Plant Proctect 10:3−12

Valleau WD (1952) Breeding tobacco for disease resistance. Econ Bot 6:69−102

Van der Plank JE (1960) Analysis of epidemics. In: Horsfall JG, Dimond AE (eds) Plant pathology, vol III. Academic Press, London New York, pp 229−289

Van der Plank JE (1963) Plant diseases: Epidemics and control. Academic Press, London New York, pp 349

Van der Plank JE (1968) Disease resistance in plants. Academic Press, London New York, pp 206

Vanderplank JE (1975a) Horizontal resistance: Six suggested projects in relation to blast disease of rice. In: Galvez GE (ed) Proc Semin Horizontal Resist Blast Dis Rice. CE-9, CIAT Cali, Colombia, pp 21−26

Vanderplank JE (1975b) Principles of plant infection. Academic Press, London New York, pp 216

Vanderplank JE (1976) Four assays. Annu Rev Phytopathol 14:1−10

Vanderplank JE (1978) Genetics and molecular basis of plant pathogenesis. Springer, Berlin Heidelberg New York, p 167

Vaneijk JP, Eikelboom W (1983) Breeding for resistance to *Fusarium oxysporum* f.sp. *tulipae* in tulip (*Tulipa* L.) 3. Genotypic evaluation of cultivars and effectiveness of pre-selection. Euphytica 32:505−510

Varma PM (1952) Studies in the relationship of the bhindi yellow vein mosaic virus and its vector, the whitefly (*Bemisia tabaci* Genn.). Ind J Agric Sci 25:293−302

Varns JL, Currier WW, Kuc J (1971) Specificity of rishitin and phytuberin accumulation by potato. Phytopathology 61:968−971

Vashistha RN, Choudhary B (1974) Inheritance of resistance to red pumkin beetle in muskmelon. SABRAO J 6:95−97

Vavilov NI (1922) The law of homologous series in variation. J Genet 12:47−89

Vavilov NI (1949−1950) The origin, variation, immunity and breeding of cultivated plants (Transl from Russian, chester). Chron Bot 13, p 364

Vavilov NI (1957) World resources of cereals, leguminous seed crops and flax. Acad Sci USSR, Moskva. Transl Paenson M, Cole ZSUS Dep Commerce, Washington DC, p 442

Vay JE de, Charudattan R, Wimalajeewa DLS (1972) Common antigenic determinants as a possible regulator of host-pathogen compatibility. Am Nat 106:185−194

Vaziri A, Keen NT, Erwin DC (1981) Correlation of medicarpin production with resistance to *Phytophthora megasperma* f.sp. *medicaginis* in alfalfa seedlings. Phytopathology 71:1235−1238

Via DJ da, Knowles PF, Klisiewicz JM (1981) Evaluation of the world sunflower collection for resistance to *Phytophthora*. Crop Sci 21:226−229

Vidyabhushanam RV (1972) Breeding for shootfly resistance in India. In: Jotwani MG, Young WR (eds) Control of sorghum shootfly. IBH, Oxford, p 324

Villareal RL (1980) The slow leaf blast infection in rice (*Oryza sativa* L.). PhD thesis, Pa State Univ USA, p 107

Villareal RL, Lantican RM (1965) The cytoplasmic inheritance of susceptibility to *Helminthosporium* leaf spot in corn. Philos Agric 49:294−300

Vir S, Grewal JS, Gupta VP (1974) Inheritance of resistance to *Ascochyta* blight in chickpea. Euphytica 24:209−211

Volk RJ, Kahn RP, Weintraub RL (1958) Silicon content of the rice plant as a factor influencing its resistance to infection by the blast fungus, *Pyricularia oryzae*. Phytopathology 48:179−184

Wallace AT, Luke HH (1961) Induced mutation rates with gamma rays at a specific locus in oats. Crop Sci 1:93–96

Wallace B (1975) Hard and soft selection revisited. Evolution 29:465–473

Wallace LE, McNeal FH, Berg MA (1974) Resistance to both *Oulema melanopus* and *Cephus cinctus* in pubescent leaved and solid-stemmed wheat selections. J Econ Entomol 67:105--107

Walter EV (1957) Corn earworm lethal factors in silks of sweetcorn. J Econ Entomol 50:105–106

Walter EV (1962) Sources of earworm resistance in sweet corn. Proc Am Soc Hortic Sci 80:485--497

Wannamaker WK (1957) The effect of plant hairiness of cotton strains on boll weevil attack. J Econ Entomol 50:418–423

Ward HM (1902) On the relations between host and parasite in bromes and their brown rust, *Puccinia dispersa* (Erikss.). Ann Bot (London) 16:233–315

Watson IA (1970a) Changes in virulence and population shifts in plant pathogens. Annu Rev Phytopathol 8:209–230

Watson IA (1970b) The utilization of wild species in the breeding of cultivated crops resistant to plant pathogens. In: Frankel OH, Bennett E (eds) Genetic resources in plants: Their exploitation and conservation. IBP Handb No 11. Blackwell, Oxford, pp 441–457

Watson IA, Luig NH (1958) Somatic hybridization in *Puccinia graminis* var. *tritici*. Proc Linn Soc NSW 83:190–195

Watson IA, Luig NH (1963) Classification of *Puccinia graminis* var. *tritici* in relation to breeding resistant varieties. Proc Linn Soc NSW 88:235–258

Watson IA, Stewart DM (1956) A comparison of the rust reaction of wheat varieties Gabo, Timstein, and Lee. Agron J 48:514–516

Watson IW, Singh D (1952) The future of rust resistant wheat in Australia. J Aust Inst Agric Sci 28:190–197

Webster JA (1975) Association of plant hairs and insect resistance. US Dep Agric Misc Publ 11297: 18

Webster JA, Smith DH, Rathke H, Cress CE (1975) Resistance to cereal leaf beetle in wheat: density and length of leaf surface pubescence in four wheat lines. Crop Sci 15:199–202

Weibe DE, Starks KJ, Wood EA, Morrison RD (1972) Sorghum cultivars and progenies rated for resistance to greenbugs. Crop Sci 12:334–336

Weiss HB (1943) Colour perception in insect. J Econ Entomol 36:1–17

Wensler RJD (1962) Mode of host selection by an aphid. Nature (London) 195:830–831

Wenzel G (1980) Protoplast techniques incorporated into applied breeding programmes. In: Ferenczy L, Farkas GL (eds) Advances in protoplast research. Pergamon, London, pp 327–340

Wenzel G (1985) Strategies in unconventional breeding for disease resistance. Annu Rev Phytopathol 23:149–172

Wessling WH (1958a) Resistance to boll weevils in mixed populations of resistance and susceptible cottons. J Econ Entomol 51:502–506

Wessling WH (1958b) Genotypic reactions to boll weevil attack in upland cotton. J Econ Entomol 51:508–512

Wheeler H, Luke HH (1955) Mass screening for disease resistant mutants in oats. Science 128:1229

Wheeler H, Williams AS, Young LD (1971) *Helminthosporium maydis* T-toxin as an indicator of resistance to southern corn leaf blight. Plant Dis Rep 55:667–671

Whittaker RH, Feeny PP (1971) Allelochemics: Chemical interactions between species. Science 171:757–770

Wiberg A (1973) Mutants of barley with induced resistance to powdery mildew. Hereditas 75: 83–100

Wickson EJ (1886) Hessian-fly and resistant grain. Calif Agric Exp Stn Bull 581

Widstrom NB, Wiseman BR (1973) Locating major genes for resistance to the corn earworm in maize inbreds. J Hered 64:83–86

Widstrom NB, Wiseman BR, McMillian WW (1972) Resistance among some maize inbreds and single crosses to fall armyworm injury. Crop Sci 12:290–292

Widstrom NW, Burton RL (1970) Artificial infestation of corn with suspensions of corn earworm eggs. J Econ Entomol 63:443–446

Widstrom NW, Hamm JJ (1969) Combining abilities and relative dominance among maize inbreds for resistance to earworm injury. Crop Sci 9:216–219

Widstrom NW, McMillian WW (1973) Genetic effects conditioning resistance to earworm in maize. Crop Sci 13:459–461

Widstrom NW, Wiser WJ, Bauman LF (1970) Recurrent selection in maize for earworm resistance. Crop Sci 10:674–676

Widstrom NW, Wiseman BR, McMillian WW (1973) Evaluation of selection potential for earworm resistance in two corn populations and their cross. Crop Sci 15:183–184

Wienhues A (1966) Transfer of rust resistance of *Agropyron* to wheat by addition, substitution and translocation. In: Proc Symp 2nd Int Wheat Genet Hereditas (Suppl) 2:328–341

Wienhues A (1967) Die Übertragung der Rostresistenz aus *Agropyron intermedium* in den Weizen durch Translokation. Zuechter 37:345–352

Wienhues A (1973) Translocations between wheat chromosomes and an *Agropyron* chromosome conditioning rust resistance. In: Sears ER, Sears LMS (eds) Proc Symp 4th Int Wheat Genet. Univ Missouri, Columbia, pp 201–207

Wilkins PW (1975) Inheritance of resistance to *Puccinia coronata* and *Rhynchosporium orthosporum* Caldwell in Italian ryegrass. Euphytica 24:191–196

Wilkins PW, Catherall PL (1977) Variation in reaction to barley yellow dwarf virus in ryegrass and its inheritance. Ann Appl Biol 85:257–263

Williams EB, Kuc J (1969) Resistance in *Malus* to *Venturia inaequalis*. Annu Rev Phytopathol 7:223–246

Williams ND, Miller JD (1982) Inheritance of resistance to stem rust in a selection of the wheat cultivar Waldron. Crop Sci 22:1175–1179

Williams ND, Miller JD, Joppa LR (1979) Inheritance of stem rust resistance in the durum wheat cultivar Ward. In: Proc Symp 5th Int Wheat Genet New Delhi 2:1057–1060

Williams PH, Walker JC, Pound GS (1968) Hybelle and Sanibel, multiple disease resistant, F₁ hybrid cabbages. Phytopathology 58:791–796

Williams RJ (1977) Identification of multiple disease resistance in cowpea. Trop Agric 54:53–59

Williams RJ, Singh SD, Pawar MN (1981) An improved field screening technique for downy mildew resistance in pearl millet. Plant Dis 65:239–241

Williams W (1964) Genetical principles and plant breeding. Blackwell, Oxford, p 504

Williams WM (1972) Laboratory screening of white clover for resistance to stem nematode. NZJ Agric Res 15:363

Wilson FD, Shaver TN (1973) Glands, gossypol content and tobacco budworm development in seedlings and floral parts of cotton. Crop Sci 13:107–110

Wilson RL, Wilson FD (1974) Laboratory diets for screening cotton for resistance to pink boll worm. Cotton Grow Rev 51:302–308

Wilson RL, Wilson FD (1975) Comparison of an X-ray and a greenboll techniques for screening cotton for resistance to pink bollworm. J Econ Entomol 68:636–638

Wilson RL, Wilson FD (1977) Effects of cottons differing in pubescence and other characters on pink bollworms in Arizona. J Econ Entomol 70:196–198

Wilson RL, Starks KJ, Pass H, Wood EA (1978) Resistance in four oat lines to two biotypes of the greenbug. J Econ Entomol 71:886–887

Wilson RL, Jarvis JL, Guthrie WD (1983) Evaluation of maize for resistance to black cut worm *larvae*. Maydica XXVIII:449–453

Wiseman BR, McMillian WW, Widstrom NW (1972) Tolerance as a mechanism of resistance in corn to the corn earworm. J Econ Entomol 65:835–837

Wit PJG de, Kodde K (1981) Induction of polyacetylenic phytoalexins in *Lycopersicon esculentum* after inoculation with *Cladosporium fulvum* (syn. *Fulvia fulva*). Physiol Plant Pathol 18:143–148

Wolfe MS, Barrott JA (1977) Population genetics of powdery mildew epidemics. NY Acad Sci 287:151–163

Wolfe MS, Schwarzbach E (1975) The use of virulence analysis in cereal mildews. Phytopathol Z 82:297–307

Wolfe MS, Barrott JA, Shattock RC, Shaw DS, Whitbread R (1976) Phenotype-phenotype analysis. Field application of gene-for-gene hypothesis in host paghogen relations. Ann Appl Biol 82:369–374

Wong LSL, McKenzie RIH, Harder DE, Martens JW (1983) The inheritance of resistance to *Puccinia coronata* and of floret characters in *Avena sterilis*. Can J Genet Cytol 25:329–335

Wood EA Jr, Chada HL, Saxena PN (1969) Reactions of small grains and grain sorghum to three greenbug biotypes. Okla Agric Exp Stn Prog Rep 618:7

Woodhead S, Padgham DE, Bernays EA (1980) Insect feeding on different sorghum cultivars in relation to cyanide and phenolic acid content. Ann Appl Biol 95:152–157

Woodsworth CW (1891) Variation in hessian-fly injury. Calif Agric Exp Stn Rep 1890:312–318

Wright DSC (1966) *Verticillium* wilt of tobacco (A note). NZJ Agric Res 9:448–451

Xu Z-H, Huang B (1984) Anther factor(s) in barley anther culture. Acta Bot Sin 26:1–10

Yadava HN, Mital RK, Singh HG (1967) Correlation studies between leaf mid-rib structure and resistance to jassids (*Empoasca devastans* Dist.) in cotton. Ind J Agric Sci 37:495–497

Yamamoto M, Matsuo K (1976) Involvement of DNA in resistance of potatoes to invasion by *Phytophthora infestans*. Nature (London) 259:63

Yamasaki Y, Kawai T (1968) Artificial induction of blast resistant mutations in rice. In: Proc Symp Rice Breed Induced Mutat. Vienna Int At Energy Ag Tech Rep Ser No 86

Yang SL, Ellingboe AH (1972) Cuticle layer as a determining factor for the formation of mature appresoria of *Erysiphe graminis* on wheat and barley. Phytopathology 62:708–714

Yarwood CE (1956) Cross protection with two fungi. Phytopathology 46:540–544

Yoder OC (1980) Toxins in pathogenesis. Annu Rev Phytopathol 18:103–139

Yoder OC, Gracen VE (1975) Segregation of pathogenicity types and host-specific toxin production in progenies of crosses between races T and O of *Helminthosporium maydis (Cochliobolus heterostrophus)*. Phytopathology 65:273–276

Yoder OC, Scheffer RP (1969) Role of toxin in early interactions of *Helminthosporium victoriae* with susceptible and resistant oat tissues. Phytopathology 59:1954–1959

Yorinori JT, Thurston HD (1975) Factors which may express general resistance in rice to *Pyricularia orzyae* Cav. In: Galvez GE (ed) Proc Semin Horizontal Resist Blast Dis Rice. Ser CE-9, CIAT Cali, Colombia, pp 117–135

York DW, Dickson MH, Abawi GS (1977) Inheritance of resistance to seed decay and pre-emergence damping-off in snap beans caused by *Pythium ultimum*. Plant Dis Rep 61:285–289

Yoshimura A, Mew TW, Khush GS, Omura T (1983) Inheritance of resistance to bacterial blight in rice cultivar Cas. 209. Phytopathology 73:1409–1412

Young HC Jr, Browder LE (1965) The North American 1965 set of supplement differential wheat varieties for identification of races of *Puccinia recondita tritici*. Plant Dis Rep 49:308–311

Yuen JE, Lorbeer JW (1983) A new gene for resistance to *Bremia lactucae*. Phytopathology 73:159–162

Zadoks JC (1961) Yellow rust on wheat: Studies in epidemiology and physiologic specialization. Tydschr Plant Ziekten 67:69–256

Zaki AI, Keen NT, Sims JJ, Erwin DC (1972) Vergosin and hemigossypol, antifungal compounds produced in cotton plants inoculated with *Verticillium albo-atrum*. Phytopathology 62:1398–1401

Zhukovskij PM (1959) Interrelation between host and parasite in their centers of origin and beyond it. Vestn Ssk Nauki (Moscow) 6:24–34

Zhukovskij PM (1961) Grundlagen der Introduktion der Pflanzen auf Resistenz gegen Krankheiten. Zuechter 31:248–253

Zhukovskij PM (1964) Cultivated plants and their wild relatives. Systematics, geography, cytogenetics, ecology, origin, and use. 2nd edn. Kolos, Leningrad, p 792 (3rd edn in Russian)

Zhukovskij PM (1965) Main gene centers of cultivated plants and wild relatives within the territory of the USSR. Euphytica 14:177–188

Zhukovskij PM (1970) World genofund of plants for breeding. Megacenters and microcenters. Manuscript (unpublished)

Zimmer DE, Schafer JE, Patterson FL (1963) Mutations for virulence in *Puccinia coronata*. Phytopathology 53:171–176

Zitelli G, Vallega J (1968) Genetical factors for resistance mainly to race 14-1 of *Puccinia graminis tritici* in some Italian varieties of *Triticum aestivum*. In: Proc Cereal Rust Conf, Oeiras, Portugal, pp 166–170

Zitelli G, Vallega J (1971) Ereditarietà della resistenza ad alcune razze di *Puccinia graminis tritici* nei frumenti teneri e introduzione di questo carattere in nuove selezioni. Sementi Elette 174:41–52

Zitelli G, Vallega V (1977) Strategies to be used in the struggle between resistance and virulence genes. In: Proc Symp Induced Mutat Against Plant Dis. Int At Energy Ag, Vienna, pp 97–107

Zohary D, Harlan JR, Vardi A (1969) The wild diploid progenitors of wheat and their breeding value. Euphytica 18:58–65

Zummo N, Broadhead DM (1984) Sources of resistance to rough leaf spot disease in sweet sorghum. Plant Dis 68:1048–1049

Subject Index

113

Printed in the United States
By Bookmasters